Micro Electro Discharge Machining

Micro and Nanomanufacturing Series

Vijay Kumar Jain
IIT Kanpur

The aim of this series is to provide the basis for knowledge transfer from research laboratory and industrial practice to the wider scientific and engineering community on the development of micro- and nanomanufacturing processes. The series will focus on the invention, development, and use of micro and nanomanufacturing processes. It will also incorporate books on new research processes, established micro- and nanomanufacturing processes, applications of existing processes to new product applications, and those focusing on case studies. This series of books should not only be popular among engineers, researchers, and professionals across a wide range of science and engineering areas but also be of interest to undergraduate- and graduate-level students as course supplemental reading.

Nanofinishing Science and Technology
Basic and Advanced Finishing and Polishing Processes
Edited by Vijay Kumar Jain

Diamond Turn Machining
Theory and Practice
R. Balasubramaniam, RamaGopal V. Sarepaka, and Sathyan Subbiah

Micro Electro Discharge Machining
Principles and Applications
Ajay M. Sidpara and Ganesh Malayath

For more information about this series, please visit: https://www.crcpress.com/Micro-and-Nanomanufacturing-Series/book-series/CRCMICANDNAN

Micro Electro Discharge Machining
Machining
Principles and Applications

Ajay M. Sidpara and Ganesh Malayath

CRC Press
Taylor & Francis Group
Boca Raton London New York

CRC Press is an imprint of the
Taylor & Francis Group, an **informa** business

CRC Press
Taylor & Francis Group
6000 Broken Sound Parkway NW, Suite 300
Boca Raton, FL 33487-2742

First issued in paperback 2020

ISBN-13: 978-1-138-61307-2 (hbk)
ISBN-13: 978-0-367-77672-5 (pbk)

Library of Congress Cataloging-in-Publication Data

Names: Sidpara, Ajay M., author. | Malayath, Ganesh, author.
Title: Micro-electro discharge machining: principles and applications / by
Ajay M. Sidpara and Ganesh Malayath.
Description: Boca Raton, FL: CRC Press/Taylor & Francis Group, 2019. |
Series: Micro and nanomanufacturing series | Summary: "Micro Electro Discharge
Machining (EDM) is a prominent technology for the fabrication of micro components in
many fields. Currently, micro EDM has a role in micro manufacturing related applications.
This book provides the fundamental knowledge of the principles of the process, the different
process parameters, the role of machine components and systems, the challenges, and how
to eliminate processing errors. An extensive study of applications in different areas with the
most relevant examples will also be included" - Provided by publisher.
Identifiers: LCCN 2019020282 | ISBN 9781138613072 (hardback: alk. paper) |
ISBN 9780429464782 (ebook)
Subjects: LCSH: Micro-electro discharge machining.
Classification: LCC TJ1191.5 .S53 2019 | DDC 671.5/212 -dc23
LC record available at https://lccn.loc.gov/2019020282

Visit the Taylor & Francis Web site at
http://www.taylorandfrancis.com

and the CRC Press Web site at
http://www.crcpress.com

Dedicated to the EDM fraternity

Contents

Preface

Nowadays, electric discharge machining (EDM) is considered as a mainstream process along with other conventional machining processes. The capability of cutting any electrically conductive material irrespective of its hardness and force-free nature of EDM has made it more acceptable to the manufacturing industries as compared to other advanced machining processes. With demand for miniature components and products, machining at microscale starts gaining attention of researchers as well as practicing engineers. Micro EDM is considered as one of the widely used machining processes for machining of small details in big parts or the machining of miniaturized components. Different variants of micro EDM extend its application domain to all sectors using microcomponents or microfeatures. Biomedical, aerospace, automotive, microelectronics, and jewelry or watchmaking industries are some of those sectors.

This book provides a comprehensive coverage of micro EDM and its variant processes. This book summarizes micro EDM process, starting from its history to the latest advancement, which is carefully reviewed and written from journal papers, conference papers, technical notes, industries catalogs, magazines, etc. This book can be used as a reference guide by students, researchers, and professionals for learning, exploring new things, selection of process parameters, tool electrodes for different workpiece materials, and many more.

This book contains ten chapters covering the most important aspects of micro EDM process. **Chapter 1** provides an overview of different micromanufacturing processes, history of EDM, working principle of micro EDM, summary of machine components as well as different variants of micro EDM. **Chapter 2** discusses micro EDM machine components such as machine tool structure, spindle system, servo control systems, pulse generators, dielectric fluids, sparking gap control systems, and tool electrodes. **Chapter 3** focuses on micro EDM milling, tool movement and rotation, and its capabilities and applications. **Chapter 4** discusses die-sinking micro EDM, machining behaviour, and capabilities and applications. **Chapter 5** focuses on micro EDM drilling, working principle, machining behavior, and capabilities and applications. **Chapter 6** discusses the different variants of micro EDM such as electro discharge grinding, reverse EDM, planetary EDM, hybrid EDM, and others related to deposition and surface modification. **Chapter 7** discusses important process parameters and their effect on the final quality of the machined surface. **Chapter 8** includes geometric error (overcut, taper, depth, etc.) and surface problems (recast layer, heat-affected zone, microcracks, residual stress, material migration, corrosion, etc.). **Chapter 9** focuses on the techniques used for the modeling of micro EDM and methodology for theoretical calculation of temperature, crater size, etc. **Chapter 10** discusses the tool wear (types and their effects) and tool wear compensation methods (theoretical as well as experimental).

Ajay M. Sidpara

Ganesh Malayath

Acknowledgments

We would like to thank Prof. V. K. Jain, series editor of *Micro and Nanomanufacturing,* for his encouragement to write this book. We also thank Taylor & Francis Group (CRC Press) in general and Ms. Cindy Carelli and Ms. Erin Harris, in particular, for providing full support during the different preparation stages of this book. Thanks to Mr. Jomy Joseph, Mr. B. Nitin, and Mr. Abhishek N. for their help in proofreading. Thanks to all the researchers, engineers, and suppliers for their scholarly publications, technical notes, and catalogs on micro EDM for improving the content of this book.

Authors

Ajay M. Sidpara completed his B.E. from Government Engineering College, Bhuj (Gujarat University) in 2001, M.E. from the Maharaja Sayajirao University of Baroda in 2004, and Ph.D. from the Indian Institute of Technology Kanpur in 2013. He is working as an assistant professor in Mechanical Engineering Department at the Indian Institute of Technology Kharagpur. He has more than 15 years of teaching and research experience. His research interests are surface finishing and micromachining.

He has received several awards (Gandhian Young Technology Innovation Award 2013, IEI Young Engineers Award 2015, and CSR Innovation Award 2016). He received research funding from BRNS—Bombay, SERB—New Delhi, and GE Power India Limited. He has published more than 30 research papers in international journals, 13 book chapters, more than 30 conference papers, and filed 3 Indian patents. He is a reviewer of more than 20 international journals.

Ganesh Malayath completed his B.Tech. in mechanical engineering from Government Engineering College, Thrissur, and M. Tech. in manufacturing technology from the National Institute of Technology, Calicut. He is currently pursuing his research on micro EDM in the Mechanical Engineering Department at the Indian Institute of Technology Kharagpur. He has published three research papers in international journals, one book chapter, four conference papers, and filed one Indian patent.

1

Introduction to Micro EDM

1.1 The Need for Micromachining Technologies and Their Applications

In the popular article written by Richard P. Feynman titled "There is plenty of room at the bottom" (Feynman, 2012), he brings up an interesting question— "Why cannot we write the entire 24 volumes of the *Encyclopedia Britannica* on the head of a pin? Followed by this, he put forward the revolutionary idea of miniaturization and the need for it. He wrote, "Many of the cells are very tiny, but they are active; they manufacture various substances; they walk around; they wiggle; and they do all kinds of marvelous things—all on a very small scale. Also, they store information. Consider the possibility that we too can make a thing very small which does what we want—that we can manufacture an object that maneuvers at that level." Today, we realize what the potential of miniaturization is. It has produced an unrealistic universe of technologies with products small in size and huge in technological capability. How are tiny components able to outperform all the advanced technologies during the period of its evolution?

- The low weight of the miniaturized components reduces the inertial effects during motion, which facilitates more precise movements of the parts.
- Tiny, weightless components are much needed in aerospace and biomedical applications.
- Smaller components require less raw material for production.
- Miniaturized components demand less space, which enables compact packaging of a more significant number of components in the system.
- Small parts can be expertly produced in batches, which reduce the cost of products considerably.
- With decrease in size, more components of different functions can be coupled to enhance the multifunctional ability of devices (Hsu, 2002).

The last couple of decades witnessed a technological boom related to miniaturization of instruments, parts, and machinery due to advancements in microfabrication techniques. As demand increases, newer technologies are introduced to the market and thrive on the goal of accurately reducing the dimensions of machine components. Most of these technologies are scaled-down versions of existing manufacturing techniques. However, the principle and challenges are not the same. Many factors that are not influential in macromachining technologies take the principal role in determining the accuracy of micromachined parts. At this scale, small changes in the processing parameters or manufacturing environment reflect on the accuracy of the final product. Microdefects in the raw material, the geometry of the cutting tool, temperature effects on the axis movements of the machine tool, and the presence of chatter or vibrations will have a proportionally high impact on the repeatability and accuracy of the micromanufacturing process.

Figure 1.1 shows some of the main application areas of microfabrication techniques. Microfabrication techniques have been proven to have a huge scope in the medical field. Figure 1.1a shows a microrobot in the size of bacteria for direct drug delivery system. Untethered microrobots are the product of advancements in fabrication techniques of microactuators and microsensors, which has helped microdevices to reach unprecedented tiny spaces in the human body (Vikram Singh and Sitti, 2016). Neurostimulation methods use an implantable pulse generator (IPG) for stimulating the brain cells, which is comprised of electrodes and their power source unit. Classic technology could only incorporate a couple of connections in the IPG unit. However, advancements in micromolding, microstamping, microforming, and microassembly techniques helped to revolutionize the technology to incorporate about 32 connections in an IPG unit with less power requirement than the conventional stimulators (Amon and Alesch, 2017). Similarly, microfabrication techniques can mass produce thermopneumatic/shape memory alloy microvalves, piezoelectric/thermopneumatic/ferrofluidic magnetic micropumps, and recirculation flow/electrokinetically driven/ droplet micromixers (Zhang et al., 2007). Microfluidic devices have become realistic and promise affordable medical diagnosis and health care. Figure 1.1f shows a lab-on-a-chip device for microfluidic applications. Microfabrication systems successfully fabricated droplet-based microfluidic circuits, which help to reduce the reagent volume by enclosing the molecular processes to the droplet volume (Gu et al., 2011). Smartphones and personal computers with advanced technical specifications are becoming cheaper and cheaper, and possess increased processing ability with the help of silicon-based micromanufacturing systems.

Microfabrication processes can be popularly classified as follows:

1. Subtractive processes
2. Additive processes

3. Deforming processes
4. Joining processes.

1.1.1 Subtractive Processes

Subtractive processes carve out the designed part features from a blank piece of a workpiece utilizing thermal, mechanical, chemical, or photon energy sources. Subtractive processes can be further classified according to the energy source they rely on for material removal. Conventional micromachining processes use the cutting tool with greater hardness to chip out material from the workpiece. Being a contact-type machining process (turning, milling, drilling), tool rigidity, tool dimensions, and the amount of contact force exerted on the microfeatures determine the minimum achievable part dimension (Huo, 2013). Other subtractive processes include laser/electron beam machining, where a high-intensity laser/electron beam is responsible for material removal (Mishra and Yadava, 2015; Volkert et al., 2007). In the photolithographic process, a photosensitive material is shaped out with the help of a light source/plasma (Chen et al., 2012). Etching processes use

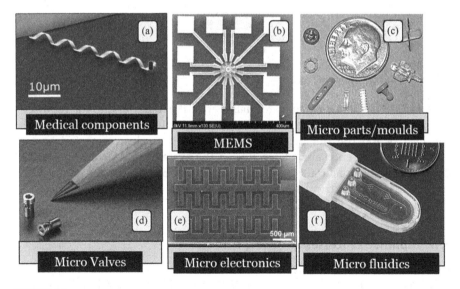

FIGURE 1.1
General areas of application of microfabrication techniques. (a) Artificial bacterial flagella—microrobots in the size of bacteria for drug delivery (courtesy: Mhanna et al., 2014; Institute of Robotics and Intelligent Systems/ETH Zurich), (b) micro electro mechanical system (MEMS) devices—a micromotor fabricated using a surface micromachining process (courtesy: www.mems-exchange.org), (c) parts fabricated by micromolding (courtesy: www.spectrumplasticsgroup.com), (d) microcheck valves fabricated by Lee Products, UK (courtesy: www.leeproducts.co.uk), (e) microsuper capacitors for supplying energy to MEMS devices (Lin et al., 2013), and (f) lab-on-a-chip component manufactured using microinjection molding (courtesy: https://microsystems.uk.com).

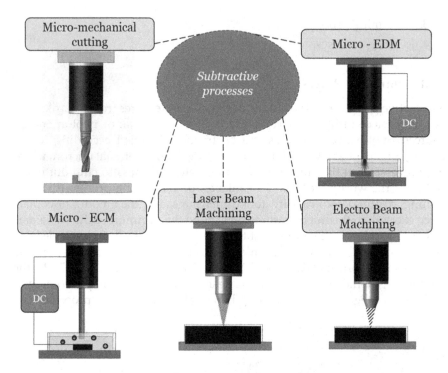

FIGURE 1.2
Subtractive microfabrication processes.

the reaction of chemical agents to disintegrate the material into the required form (Williams et al., 2003). Electrochemical machining (ECM) process uses the idea of ion transfer between conductive electrodes connected across a potential difference and electrolyte in the system for machining electrically conductive materials (Sen and Shan, 2005). Electro discharge machining (EDM) uses electrical sparks to melt and evaporate workpiece material (Pham et al., 2004). Figure 1.2 shows an overview of the subtractive processes.

1.1.2 Additive Processes

In additive processes, the microparts are manufactured by layerwise addition of material (Vaezi et al., 2013). It can be further classified into scalable additive manufacturing process (stereolithography, selective laser sintering, inkjet printing process, fused deposition modeling), 3D direct writing (precision pump and syringe-based methods, laser-induced forward transfer, beam-based deposition), and hybrid processes (electrochemical fabrication, shape deposition modeling). Selective laser sintering uses a laser beam for melting the layers of metal powder in e desired manner to produce the final part (Vaezi et al., 2013). Electrochemical fabrication process makes the microparts by structural material deposition, sacrificial material

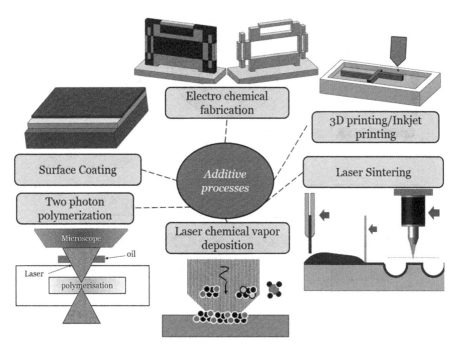

FIGURE 1.3
Additive microfabrication processes.

deposition, and planarization (Vaezi et al., 2013). Even though direct writing techniques are primarily developed for 2D structuring applications, laser chemical vapor deposition methods can also be used for fabrication of 3D microstructures. Figure 1.3 shows the additive microfabrication processes.

1.1.3 Deforming Processes

Deforming processes such as microstamping, deep drawing, extrusion, forging, and incremental forming have the advantage of less material wastage and negligible thermal effects (Razali and Qin, 2013). However, the formability of material has a major role in determining the time and feasibility of fabrication. Injection molding and casting are the next most popular microfabrication processes in which the molten material is allowed to solidify inside a mold to fabricate complex microfeatures (Heckele and Schomburg, 2004). Figure 1.4 shows some of the forming microfabrication processes, and Figure 1.5 shows the common variants of casting/molding techniques. Microjoining processes such as solid-state bonding, soldering, fusion welding, and adhesive bonding allow us to assemble different microparts without damaging the parts individually and as an assembly (Jain et al., 2014).

In all these processes, controlling the machining variables, machine tool, work, and tool holding, cutting tools or workpiece will have a cumulative

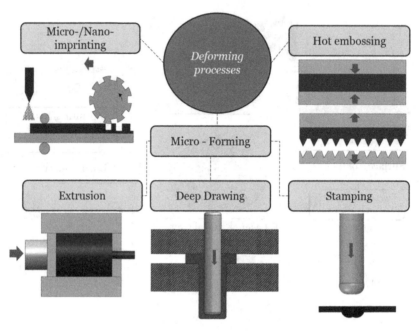

FIGURE 1.4
Deformation-based microfabrication processes.

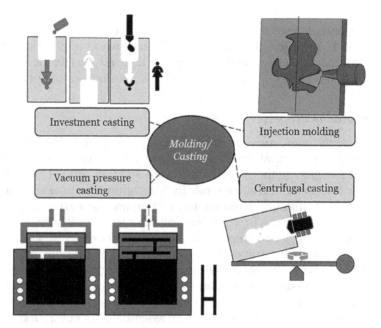

FIGURE 1.5
Casting/molding-based microfabrication processes.

effect on the result. As the size of the part reduces, accuracy gains a whole new meaning. A tolerance of ±50 µm will suffice in machining a part of 5 mm × 5 mm cross section. However, for a part with 100 µm × 100 µm cross section, this will be a huge dimensional error. Handling of smaller parts is also a tremendous challenge. The rigidity of microparts and their features are very less. Rigidity issue causes handling-based shape distortions to the final part. Furthermore, as the part size reduces, it demands complex tooling and fixture arrangements. To summarize, there are several key areas of concern when machining details are this small. Machining environment, reproducibility, and repeatability of the machine tool; effect of machine/tool vibration; handling of microparts before and after machining; management of cutting fluids/lubricants; and resolution/precision of machine tool movements all become much more critical to the success of the production of microparts.

Lithographic processes often fail to fabricate complex 3D surfaces as the material is removed layer by layer in a 2D plane. Moreover, the applicability of these processes for mass production in the manufacturing industry is challenging as the material removal rate (MRR) is low, and it is more challenging to fabricate structures with high aspect ratio. Photolithographic processes are capable of making microfeatures with high resolution. However, there is a limitation in the range of materials that can be processed using this technique (Takahata et al., 2000). Moreover, a high capital cost is needed to implement these technologies (high-cost machine tools and expensive supporting technologies are required for mask fabrication and for maintaining clean room environment). Although MRR is high for laser beam machining, it lacks in surface quality and dimensional accuracy (Rasheed, 2013). However, the conventional machining processes including milling, drilling, and turning offer fabrication of 3D structures with better processing time and a large range of materials (Huo, 2013). Nonetheless, unavailability of adequate cutting tools with high rigidity limits the use of conventional machining processes in certain directions. As the size reduces to microdimensions, the rigidity of the cutting tool reduces considerably. This leads to early breakage of the cutting tool. Frequent tool changes are unavoidable while machining difficult-to-cut materials. As the microcutting tools fabrication process is very sophisticated, cutting tool cost is very high when the dimension is less than 200 µm. So, the total machining cost will be very high because of frequent tool breakage during machining. Additive manufacturing processes promise rapid production of microparts without wasting much raw material such as subtractive processes. However, the process is very slow, and the surface quality of the final product demands some postprocessing to make it compatible for end use. The usable material range also has limitations.

Electrophysical/chemical micromachining processes put forward an interesting solution for most of these problems. Micro EDM and micro ECM are capable of fabricating 3D structures with great accuracy with the help of computer numerical control (CNC) axes movements (Jahan et al., 2014). As the tool responsible for material removal has no direct contact with the workpiece surface, the problem of tool breakage never arises. Material removal mechanism responsible for machining using ECM and EDM does not have any dependency on the hardness of the workpiece. Micro ECM provides highly finished surfaces, but the machining time has to be compromised as the rate of material removal is very low. On account of this, micro ECM is popular in the areas where highly finished complex 3D structures are to be fabricated irrespective of the long machining time. However, micro EDM is an electrothermal process in which melting and vaporization are responsible for material removal. The MRR is comparatively high, and it is very much effective in producing complex 3D surfaces.

1.2 Micro EDM as a Prominent Micromachining Technique

One of the main characteristics of EDM that gives a prominent role in the microfabrication industry is the force-free nature of the process. The tool and the workpiece are always separated by a distance called "spark gap." This noncontact behavior of machining enables micro EDM process to make very small microparts without making shape distortions to the microfeatures due to external force. Furthermore, it demands comparatively low capital investment. Micro EDM gives high design freedom for the manufacturer as it demands simple fixturing and small setup time. Chance of chatter or vibrations during machining is also minimum (Pham et al., 2004). The effect of vibration, thermal stress, and heat-affected zone is considered to be negligible in the case of micro EDM. Micro EDM can be used to machine a large variety of materials with minimum conductivity ($k > 0.01\,S/cm$) (Qin, 2010).

1.2.1 History of EDM

In a letter to John Canton in June 1766, Joseph Priestley wrote that "... the metal first melted, then being liquid is thrown forward or rises as water would do, sometimes making bubbles that then burst to give a crater-like appearance, deeper in gold than in silver" (Priestley, 1769). He was amazed by the strange behavior of sparks eroding metals as it strikes on the surface, creating some odd circular rings on the surface around the central crater known as Priestley's rings. Figure 1.6 shows an anodic crater formed due to an arc discharge. Priestley used Leyden jar batteries as the high-voltage source, which was not capable of providing energy supply for a long time and resulted in intermittent discharges. Even though it was a disadvantage using Leyden jar batteries, it became the base stone of different technologies

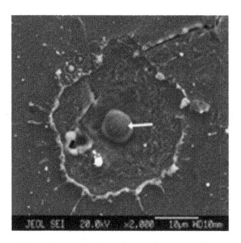

FIGURE 1.6
A crater formed with the action of an arc discharge (Zhang et al., 2014).

known today (explosive emissions, the formation of erosion crater, ability to form microparticles, and ability to coat one material over other). Priestley further observed that the depth of crater also depends on the electrode material. As an introduction to the writings on the origin of plasma science, Andre Anders observes Priestley's rings as the phenomena associated with damped oscillations of the electrical circuitry. He observed, "We know that a higher discharge current causes the number of arc spots operating simultaneously to increase rather than a change in the character of individual spots. The number of spots or current per spot also depends on the material and its surface conditions" (Anders, 2003).

Before Presley's experiments, Robert Boyle used electrical discharges to make powder from metal rods in the 17th century. Arc discharges were used to weld the metals in 1881 by Meritens. Thomas Edison exploited the idea of cathode arcs for plating metals on different surfaces. First, a US patent was filed to machine diamond and steel using electrical discharges in 1930 (Schumacher et al., 2013). In the same time, electrical discharges were used for deburring operation for the first time by a US company called Elox. However, the applicability of electrical discharges for machining was realized only by the mid-20th century. One of the main hindrances in using electrical discharges for shaping metals is the unpredictability and nonuniform behavior of the discharges. Compared to arc discharges used for welding, discharges used for machining metals have to be more uniform and capable of constantly removing a precise amount of material per spark. In welding, it is desirable to keep the melt pool on the surface of the workpiece for joining two different parts. However, in machining, the melt pool has to be thrown out from the cavities somehow. One of the major developments towards controlled machining using electrical discharges happened during the Second World War in USSR.

At that time, tungsten was known to be one of the hardest materials, so it was chosen as the material for contact points in distributor circuits of military vehicles to avoid erosion due to friction. However, this advantage was shadowed by the occurrence of small discharges at the contact points. These discharges adversely affected the quality of contact points as small pits appeared on the surface due to material erosion from discharges. The contact point failed to perform efficiently and led to frequent replacements of these tungsten points. Breakdown of military vehicles became a constant headache for the army, and changing the tungsten contact point was an expensive maintenance routine.

A scientist couple, Dr Boris Lazarenko and Dr Natalya Lazarenko, in USSR was asked to find a solution to this problem. They desperately tried many techniques to solve this issue. As part of their trials, they immersed the contact points in some mineral oils (Jameson, 2001). Even though the mineral oils were not capable of vanishing the sparks between the contact points, they observed some interesting phenomena. Mineral oils can some-how make the sparks more uniform and predictable. Uniform sparks led to more uniform pitting over the tungsten surface, one of the hardest materials to be known as said before. Figure 1.7 shows the experimental arrangement of Dr Boris Lazarenko and Dr Natalya Lazarenko with the spark ignition system. The Lazarenko couple immediately realized the potential of this invention, and they decided to make a machining system based on their observation, which later became one of the popular names in the manufacturing industry. In their thesis titled "Inversion of the Effect of Wear on Electric Power Contacts for Machining Purposes," the Lazarenko couple presented their prototype of EDM machine with a structure similar to a spark distributor unit. Figure 1.8 shows Dr Boris Lazarenko and

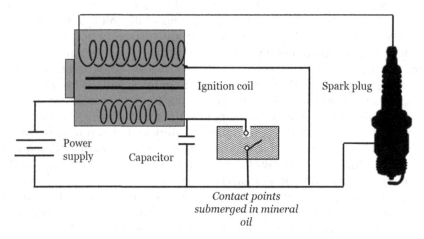

FIGURE 1.7
Lazarenkos' experiment with contact breaker points immersed in mineral oil.

FIGURE 1.8
Photograph of Dr Boris Lazarenko and Dr Natalya Lazarenko working on the prototype of the first EDM machine (Schumacher et al., 2013).

Dr Natalya Lazarenko and their EDM prototype. Figure 1.9 shows the schematic diagram of the first EDM prototype invented by the Lazarenkos.

The Lazarenko couple further developed the system by introducing a servo control system for keeping the spark gap constant during machining. The Lazarenko EDM machine designs are then adopted by Japan and the UK. The development of EDM in the USA took a completely different path.

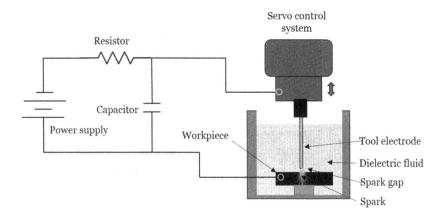

FIGURE 1.9
Lazarenkos' EDM machine prototype.

Breakage of drill bits and taps is a common thing to happen during machining. The removal of jammed taps and drills from the hole was a tedious work. Three engineers Harold Stark, Victor Harding, and Jack Beaver, who worked in an American valve-making company, were asked to find an easy way to remove broken taps and drills from the hole (Jameson, 2001). With the knowledge in electrical engineering, Harding was the one who suggested the use of electrical sparks for disintegrating the broken pieces of tools from the hole. As the sparks successfully melted parts of the tools, they were able to remove the broken pieces from the hole. Figure 1.10 shows a schematic diagram of the machining experiment by H. Stark, V. Harding, and J. Beaver (Jameson, 2001). However, this process was frustratingly slow to present as an industrial solution for a company producing valves in mass. Thus, they tried to increase the spark intensity. As the power of the sparks increased, the erosion time reduced substantially. However, the surface quality of the product was affected by leftover molten layers of materials on the workpiece surface.

To remove the molten metal from the surface during machining, they introduced water as a coolant. They also introduced a vacuum tube-based pulse generating system for power supply. Vacuum tube pulse generators were successful in increasing the spark frequency by several folds. They

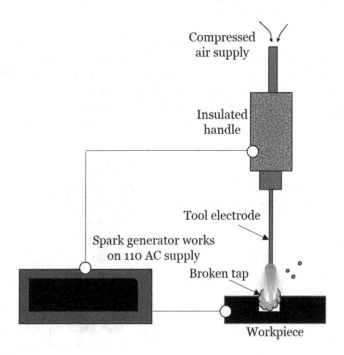

FIGURE 1.10
Schematic representation of an experimental setup for removing broken taps from holes developed by H. Stark, V. Harding, and J. Beaver.

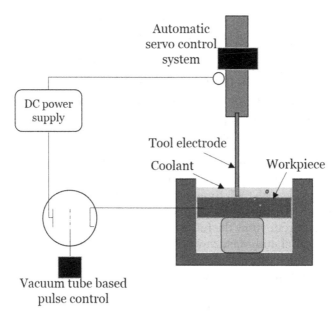

FIGURE 1.11
Prototype of the first vacuum tube-based EDM machine.

came up with an electronic circuit-based servo control system to precisely control the spark gap. These events led to the invention of vacuum tube-based EDM machines. The vacuum tube-based EDM technologies further evolved to transistor-based circuits after the invention of planar transistors. Figure 1.11 shows the schematic arrangement of a vacuum tube-based EDM machine tool (Jameson, 2001).

The technologies related to EDM have then witnessed an evolutionary boom. CNC-based axes control systems provided more precision in part dimensions, tool erosion compensation systems reduced the shape distortions in part geometry, new dielectric fluids were introduced, and the pulse generators were frequently modified for better performance. However, the theories on metal removal mechanism of EDM were ambiguous and weak. So, scientists further focused their work to theoretically understand the process. The 1950s and 1960s witnessed more and more hypotheses to explain spark formation, material removal mechanism, mechanism of spark gap control, and thermal distortions of electrodes. One of the main drawbacks of the resistor-capacitor (RC) circuit used for pulse generation is the low discharge rate due to the delay in charging of the capacitor in the circuit. Moreover, the pulse ON–OFF time cannot be controlled in an RC pulse generator. Scientists used RCL circuits that promised higher discharge rate. However, the stochastic nature of discharges increased due to the presence of an inductor in the circuit, and the discharge energy became nonuniform in each cycle. EDM evolved to an advanced machining technique with the following inventions in the late 1950s and 1960s.

- Invention of a planar transistor by Hoerni and Noyce in 1959 led to the invention of static pulse generators
- Development of servomechanisms based on integrated circuits
- Introduction of a microprocessor by Intel in 1971 led to the development of CNC systems later.

Transistor-based pulse generator systems gained popularity due to their high discharge frequency and the flexibility to control pulse characteristics by changing pulse ON–OFF time. Finally, EDM was popularly adopted by the industries. Even though EDM machines based on RC pulse generators continued their production, the future of this technology was in doubt till the late 1960s. Kurafuji and Masuzawa (1968) drilled some microholes of 6 and 9 μm in 50-μm-thick carbide plates. They also machined rectangular microholes of 40 μm × 45 μm cross section on the same material using die-sinking EDM. Instead of transistor-based circuits for pulse generation, they relied upon RC circuits for their experiments. They observed that RC circuits are capable of reducing the discharge energy level to very low values with very small pulse duration so that the material can be precisely removed from the workpiece. In transistor-based circuits, the switching component and pulse control circuitry always experience a delay in transmission so that keeping very small pulse duration cannot be possible. In RC-based circuits, the capacitor discharges very quickly, which helps to achieve discharge duration in the order of nanoseconds.

Moreover, using a capacitor with very low rating (in the order of picofarads), an ultrasmall discharge energy can be realized. These experiments kept RC circuits alive and led to a revolutionary idea of precision micromachining using electrical discharges, popularly known as micro EDM. Even though the invention of micro EDM happened in 1968, it took a long time to understand the process capabilities, optimize the process parameters, and extend its areas of application. One of the main challenges was to fabricate the thin tool electrodes. Masuzawa et al. (1985) proposed the idea of wire electrodischarge grinding (WEDG). In this, a wire of very small diameter can be used to reduce the diameter of an electrode down to 10 μm. This invention helped to fabricate EDM tool electrodes quickly and effectively. The invention of an electrode feeding device using an impact drive mechanism (IDM) helped to pull back the electrode quickly during short circuits (Furutani et al., 1997).

Masuzawa et al. (2002) developed an EDM lathe system with different micromachining capabilities, including fabrication of coaxial microparts. Micro EDM was becoming more and more popular, as novel ideas were flowing to enhance its technical capabilities. However, tool electrode wear during machining was proportionally high in micro EDM compared to conventional EDM, as the tool electrode size is very small. Characteristics of electrode wear in micro EDM are systematically analyzed by Tsai and Masuzawa (2004) by dividing the volumetric wear into two parts, namely linear wear (change in length) and corner wear (change in shape). They reported that materials with a high boiling point, thermal conductivity, and melting point are the best choices for tool electrode in micro EDM, and the volumetric wear ratio decreases with discharge energy. According to them, the wear phenomena cannot be eliminated completely, but the relative wear can be reduced below 1% by carefully selecting the machining parameters. Even though micro EDM milling is considered to be the simplest of all the EDM micromachining strategies, because of tool wear issues, it was less prevalent in the industries. Yu et al. (1998) developed an ingenious method to tackle the tool wear problem in milling. They suggested layer-by-layer machining in a specified tool path with frequent vertical tool feed, which makes an effective tool wear near to zero.

Allen et al. (1999) developed a two-stage micro EDM setup and successfully fabricated microholes in inkjet printing nozzles. Microstructuring of monocrystalline silicon was then reported (Reynaerts and Van Brussel, 1997). Morgan et al. (2003) analyzed the accuracy of electrodes machined with WEDG by fabricating 81 microshafts with an average diameter of 50 μm. Efficient flushing was also considered as an untamable problem in micro EDM. A planetary movement of the tool electrode for enhancing the dielectric flow was suggested as a solution (Egashira et al., 2006). Because of the invention of CNC machines and computer aided design and manufacturing (CAD/CAM) systems, most of the machining strategies were on the way of automating the manufacturing process. CAD/CAM systems helped to convert the part model to CNC codes and optimized the tool paths for minimum machining time. However, in micro EDM tool, electrode wear became the main hindrance in realizing an efficient CAD/CAM system for machining. Rajurkar and Yu (2000) developed an integrated CAD/CAM system by incorporating the uniform wear method for tool electrode wear compensation and successfully machined 3D surfaces with the proposed technology. CAD/CAM systems for micro EDM are further developed by Meeusen et al. (2002), Zhao et al. (2004), Tong et al. (2007), among others.

As the micro EDM technology was advancing at a fast pace, researchers were desperately trying to apply it in different areas to extend its capability. Ali and Hung (2000) used micro EDM to make micromold inserts with nickel beryllium alloy and compared it with the LIGA process for performance analysis. They have reported a surface finish (R_a) of 140 nm and a maximum aspect ratio of 6. Employing micro EDM as a micromachining technique reduces the fabrication cost by up to 20% compared to the LIGA processes. Yu et al. (1998) tried to machine noncircular holes with micro EDM. Despite its popularity as a subtractive process, micro EDM was used to fabricate a microrod of 0.14 mm diameter and 2.2 mm height using material deposition. Another interesting way of using micro EDM was for alloying metals (Ori et al., 2004). Takezawa et al. (2004) observed the formation of needlelike structures (with a tip radius of 100 nm) when 0.1 mm tungsten electrode was subjected to single discharge machining for a few hundreds of microseconds. They utilized this phenomenon for fabricating very thin microelectrodes. By this time, the microelectrode fabrication process was stabilized with the introduction of various techniques such as sacrificial block EDG, moving block EDG, disc EDG, etc. One of the interesting methods among them is the fabrication of microelectrodes by self-drilled holes (Yamazaki et al., 2004). In this, an electrode is used to drill a hole in a workpiece, and the same hole is utilized for processing the electrode by switching the polarity and offsetting the axis. A new high-speed electrode fabrication system comprised of twin wire was also developed (Sheu, 2008). Scientists were on a pursuit to machine the micropart with minimum achievable dimensions on a variety of workpiece materials. A series of microdiscs is used as a tool electrode to cut multiple microslits of 10 μm width (Kuo and Huang, 2004). Three-dimensional microcavities are machined on a polycrystalline diamond (PCD) with the help of micro EDM scanning (Sheu and Cheng, 2013). In 2013, a stylus was precisely assembled to the tactical head of coordinate measuring machine (CMM) using micro EDM (Sheu and Cheng, 2013).

To increase the precision of micro EDM, the spark generation systems have to be modified effectively. Some of the researchers dedicated their experiments towards this. The effects of discharge energy with shorter durations (<200 ns) and their thermal effects were scientifically analyzed (Wollenberg et al., 2001). Using different simulation models, they found out a method to calculate the maximum discharge energy that can be used for a particular material and tool geometry, without making short circuits and arc discharges. RC pulse generators generally lack in providing good MRR and uniform surface finish due to varying discharge energy. Frequent thermal damage happens to the workpiece and tool surface because of nonrecovery of dielectric strength between two pulse cycles. An isopulse generator, a modified version of a transistor pulse generator, could shape metals 24 times faster than conventional RC circuits (Han et al., 2004). A capacity-coupled pulse generator is realized to reduce the effects of stray capacitance in the RC circuit (Kunieda et al., 2007). An advanced servo control system that ensures the synchronized control of the piezotable and Z-axis improved the response characteristics of the servo system (Kunieda et al., 2007). Apart from the use of constant polarity electrodes for machining, a polarity changing technique for micro EDM is suggested (Pradhan and Bhattacharyya, 2008). They reported an improvement in the machining efficiency with the varying polarity technique when the pulse ON time is below 10 ms and the peak current is 1 A.

The possibility of using bipolar pulses for machining has been studied (Chung et al., 2007; Yang et al., 2018). A three-level pulse width modulation (PWM) power amplifier based on a field-programmable logic gate array is used to improve the conventional power amplification techniques (Guo et al., 2016). Today, micro EDM is used for precise machining of microparts for almost all areas, including aerospace, biomedical, automobiles, optics, and electronics industry. With the help of new tool electrode materials, new power generators, advanced tool wear compensation techniques, effective dielectric flushing techniques, advanced machine vision systems, CAD/CAM systems, and servo control systems, micro EDM is known to be one of the prominent technologies in the current manufacturing regime.

1.2.2 The General Principle of Micro EDM

Micro EDM makes use of the phenomena of spark generation when two electrically conducting electrodes are brought together under a potential difference. When more and more sparks are generated, considerable material removal from both the electrodes takes place. When two electrodes are connected across a potential difference, as shown in Figure 1.12, the electrons from the cathode experience a tendency to move towards the anode. When the electrons overcome the work potential, the cold emission of

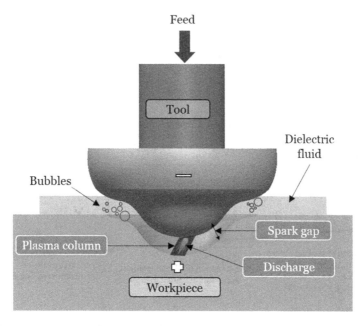

FIGURE 1.12
General principle of EDM.

electrons from the cathode surface occurs. These electrons travel towards the positively charged electrode but encounter with the dielectric fluid in between. These collisions led to the dielectric breakdown, i.e., the dielectric molecules split to electrons and ions, which reduce the effective resistance and increase the conductivity of the dielectric fluid. This transition from an insulating environment to conducting environment between the two electrodes causes a continuous flow of electrons. This avalanche of electrons collides with more molecules of dielectric fluid, which leads to the generation of a more number of electrons and ions. Vaporization of the dielectric fluid results in the formation and expansion of bubbles, which covers the region and reduces the effective resistance (Jameson, 2001). All these events facilitate the formation of a plasma channel (a material state comprised of electrons and ions), and the electron flow channelizes to the point where the spark gap is minimum (as the resistance will be minimal at this point) (Kunieda et al., 2005) as shown in Figure 1.13. This continuous flow of electrons is known as a discharge (similar to discharge in a river) (Llanes, 2001). The electron bombardment with the surface of the workpiece results in the transformation of kinetic energy to thermal energy. This sudden transfer of thermal energy melts the workpiece material locally, and some part of it evaporates instantly. The ions that travel towards the cathode surface constitute the erosion of material from tool surface, popularly known as tool wear. As the mass of ions is greater than the mass of electrons, ions are expected to remove more material. Surprisingly, if the same material is used

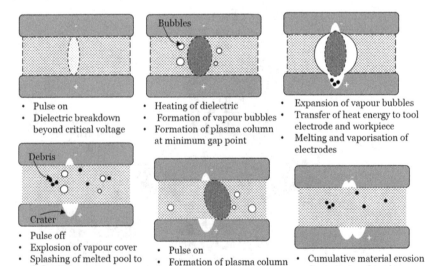

- Pulse on
- Dielectric breakdown beyond critical voltage

- Heating of dielectric
- Formation of vapour bubbles
- Formation of plasma column at minimum gap point

- Expansion of vapour bubbles
- Transfer of heat energy to tool electrode and workpiece
- Melting and vaporisation of electrodes

- Pulse off
- Explosion of vapour cover
- Splashing of melted pool to dielectric
- Regaining dielectric strength

- Pulse on
- Formation of plasma column at next minimum gap point

- Cumulative material erosion due to series of sparks

FIGURE 1.13
Graphical explanation of spark formation and material erosion in micro EDM.

as a tool and workpiece in EDM, then the electrode that connects to the positive polarity (anode) experiences more material removal compared to the cathode. This odd phenomenon can be explained by the following facts (Ghosh and Mallik, 1986):

- As the mass of the particle increases, inertial effects will be dominant, and the acceleration experienced by the ions towards cathode will be comparatively low. Less number of ions reach the cathode surface with lower kinetic energy compared to the tremendous number of high-velocity electrons hitting the anode surface within the specific pulse duration.
- The hydrocarbon oil deposits a thin carbon layer over the cathode surface.

During micro EDM, tool electrode is connected to cathode and workpiece electrode to anode. As the supply turns off after pulse ON time, the spark disappears, plasma channel gets destroyed, and the vapor covering explodes, creating a vacuum where the new dielectric fluid molecules will rush to the spark gap. Fluid in the spark gap regains its dielectric strength. The pressurized molten pool of material splashes to the dielectric fluid and washes away. The new cycle of events starts with the starting of the next pulse ON time, which leads to the generation of next spark. Because of material removal from the workpiece and tool electrode, the "minimum gap point" changes to another location as shown in Figure 1.13. The plasma channel formed at this new location leads to the formation of a crater on the specific area. Compared to the arc discharges, the pulsed power supply ensures uniform distribution of discharges over the surface, enabling uniform material removal from the workpiece surface.

1.2.3 Micro EDM vs. Macro EDM

Micro EDM can be considered as the scaled-down version of EDM. In both processes, material removal is taking place due to the ablation of material by melting and vaporization. However, to remove material in a precise manner, we have to design the system to restrict the energy intensity to a lower range. Therefore, the differences in plasma radius and energy intensity are recognized as the main differences between micro and macro EDMs. Furthermore, the material available to dissipate the thermal energy is very less in the case of microparts. Excessive spark energy may burn the tool electrode, and excessive flushing pressure will cause tool breakage or deflection. The electrode wear rate is proportionally higher in micro EDM, which demands highly sophisticated tool wear compensation methods (Jahan et al., 2014). Empirical relations that predict the trend of MRR or tool wear rate in macro EDM are not valid in the microregime. Considering these facts, micro EDM has to be treated as a separate machining strategy to understand the mechanism behind material removal and factors affecting the machining quality.

1.2.3.1 Power Supply

Macro EDM: In macro EDM, the main aim is to obtain a higher MRR with acceptable dimensional accuracy and surface finish. Theoretically, there is no limitation to the discharge energy that can be used for machining in EDM. However, to obtain adequate surface quality, extreme voltage and current conditions are not preferred. Transistor-based pulse generators are popularly used to generate sparks because of the higher frequency of sparks and flexibility in setting the pulse duration levels. Pulse duration is in the order of milliseconds. Peak current is greater than 3 A, and the source voltage has a larger range of 4–400 V (Raju and Hiremath, 2016).

Micro EDM: In micro EDM, the main aim is to obtain precise micro-features with a great deal of dimensional accuracy. Micro EDM demands discharge energy in the order of microjoules and pulse duration in the range of 50 ns to 100 μs (Raju and Hiremath, 2016). For this, the power supply has to be modified, which can give very low discharge energy for the ultrasmall duration of time. Delay in transistor-based circuits makes them not suitable to achieve small and uniform pulse durations. RC-based circuits with very low capacitor ratings can be used to produce small discharge energy. As the capacitor discharging time is very small, pulse durations in the order of nanoseconds can also be achieved. Voltage in the range of 10–120 V and peak current less than 3 A are preferred in the case of micro EDM. The upper limit of discharge energy must be carefully chosen so that the tool electrode does not get burned. The frequency of pulses has to be high for good surface finish and surface uniformity (Liu et al., 2010).

1.2.3.2 Tool Size/Spark Gap/Plasma Radius

Macro EDM: Tools of dimensions in the order of millimeters are used for machining in conventional EDM. The tool electrode is more rigid and less susceptible to thermal distortions or burning. The tool electrode is less affected by the flushing pressure, and internal flushing can be easily employed because of the higher dimensions of the tool. The plasma radius in macro EDM is much less than the diameter of the tool, and the amount of tool erosion compared to the total volume of the tool is low. The spark gap is in the range of 10–500 μm (Raju and Hiremath, 2016). The tool electrodes can be fabricated using any other machining process and transported to the EDM machine tool.

Micro EDM: Microelectrodes of dimensions less than 1 mm are chosen as the tool in micro EDM. These tools are less rigid and susceptible to bending during transportation, handling, or due to excessive flushing pressure. Because of this, the flushing pressure has to be moderated according to the tool material. In addition to that, on-machine fabrication of the tool electrode is recommended to avoid bending, breakage, and tool mounting-related problems. The plasma radius is considered to be comparable to the tool size. Spark gap has to be maintained below 10 μm (Raju and Hiremath, 2016).

1.2.3.3 Crater Size/MRR/Surface Roughness/ Minimum Achievable Size of the Part

Macro EDM: As the discharge energy is higher in macro EDM, it produces larger craters. As the crater size increases, the surface roughness is increased, and the MRR will be higher. Figure 1.14 shows a comparison of crater size in macro and micro EDMs (Uhlmann et al., 2005). The surface created by EDM has a roughness Ra in the range of 3–30 μm. MRR is in the range of 0.02–0.8 mm³/s (Rajurkar et al., 2006). Minimum achievable dimensions are in the order of millimeters.

Micro EDM: Micro EDM uses very small discharge energy, which produces crater volume in the order of cubic micrometers. This reflects on the surface quality, as the roughness is in the range of 0.05–0.5 μm (Liu et al., 2010). Discharge frequency of RC circuits is comparatively lower than that of transistor circuits. This also makes the machining process slow (MRR = 200–1,000 μm³/s). Even though the minimum achievable size is different for different variants of micro EDM, all processes are capable of fabricating features of dimension less than 50 μm (Liu et al., 2010). Microwire EDG could machine microcutting tools of dimension down to 3 μm (Egashira et al., 2011). Table 1.1 summarizes the differences between conventional (macro) and micro EDMs.

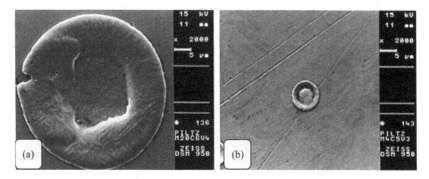

FIGURE 1.14
The difference in size of the crater produced by (a) macro EDM and (b) micro EDM (Uhlmann et al., 2005).

TABLE 1.1

Summary of the Differences between Micro and Macro EDMs

Type	Voltage (V)		Current (A)	MRR (mm³/s)	Surface Quality R_a (μm)	Interelectrode Gap (μm)	Specific Energy
EDM	Milling	20–200	20–300	0.002–0.5	3–25	10–500	High
	Die Sinking	30–300	20–400	0.002–0.8	2.5–30		
	Wire	30–120	0.5–128	5–15	0.127–0.254		
Micro EDM	10–120		<3	0.2×10^{-6} to 1×10^{-6}	0.05–0.5	<10	Low

Source: Liu et al. (2010) and Rajurkar et al. (2006).

1.3 General Components of Micro EDM

Micro EDM machine tool consists of the power supply, pulse generator, tool electrode, spindle system, worktable, CNC axes with independent and precise axis control, servo control system, dielectric circulation unit, and measurement systems. Figure 1.15 shows an overview of the micro EDM machine tool components.

1.3.1 Spark Generator/Pulse Generator

Pulse generators are responsible for making pulsed discharges for EDM. Pulse generators are of two kinds: transistor based and RC based.

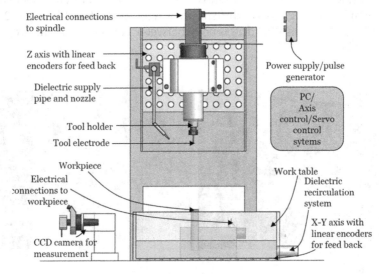

FIGURE 1.15

A schematic representation of micro EDM machine components.

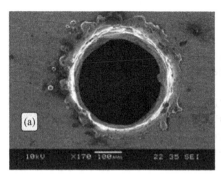

FIGURE 1.16
Entrance section of the 300 μm microhole drilled with (a) transistor-based pulse generator (140 V, 6.8 Ω) and (b) RC-based pulse generator (140 V, 4,700 pF) (Jahan et al., 2009).

Among this, RC circuit is the most popular spark generation system in which very short pulses of nanosecond duration can be realized. The discharge energy can be brought to ultrasmall levels using capacitors of very low rating.

Transistor pulse generators use a field-effect transistor for a pulse ON and OFF control. The pulse duration can be easily controlled using a transistor-based circuit. Even though RC circuits are popular in micromachining, the effect of using transistor-based circuit is studied by Masuzawa (2000), and it is reported that the generation of short pulses is only capable of rough machining. According to Han et al. (2004), the delay time in transistor-based circuits sometimes exceeds the pulse duration time, which resulted in non-uniform sparks. Figure 1.16 compares the microholes drilled using RC pulse generator and transistor pulse generator, in which a burr-like recast layer appeared around the hole in transistor-based pulse generator (Jahan et al., 2009). Han et al. (2004) developed an isopulse generator as a modified version of transistor circuits for fine machining, which has an MRR 24 times that of machines with RC circuits.

1.3.2 Servo Control System

The spark gap in micro EDM has to be continuously monitored and maintained. A servo control system is an important part of micro EDM, which avoids the occurrences of short circuits and arc formations by pulling back the tool electrode time to time to the optimum spark gap. Servo control system makes use of the ignition delay time to control the spark gap. A microcontroller takes control over the servomotor with the help of an analogue to digital (A/D) converter, a PWM module, and a feedback system (Yang et al., 2012). It monitors the discharge pulses continuously to predict the optimum spark gap based on certain algorithms for minimum arc formation and short circuits.

1.3.3 The Dielectric Fluid Supply System

The dielectric fluid is the essential part of micro EDM responsible for predictable and uniform sparks, debris removal, and cooling. The choice of dielectric fluids also changed from time to time. Experiments are conducted using deionized water (Chung et al., 2007), kerosene (Chow et al., 1999), mixed powder fluids (Jahan et al., 2011), plasma jet (Yan et al., 2018), and cryogenic coolants (Manivannan and Pradeep Kumar, 2018).

The dielectric fluid delivery system consists of a tank, pump, filters, pipes, and nozzle. The flow rate can be adjusted with valves, and the fluid is constantly filtered and recirculated. The presence of debris affects the effective resistance of the dielectric fluid. So, filters and dielectric fluid have to be changed frequently.

1.3.4 Positioning System

While fabricating features of microdimensions, the resolution and accuracy of the axes have to be very high. The axis control system in micro EDM comprises servomotors for movement and linear encoders (with nanometer resolution) for feedback. The slide straightness has to be maintained near to $2\,\mu m/100\,mm$. The base must have high stiffness and damping abilities for vibration isolation.

1.3.5 Measurement System

An online measurement system has to be incorporated in the machine tool for monitoring and measuring the microfeatures. It helps to immediately find out the irregularities in the fabricated part without removing it from the spindle or vice. A charge-coupled device (CCD) camera is popularly used for measuring the microdimensions of on-machine fabricated microtools and microparts.

1.4 Machining Parameters/Performance Parameters

Performance of micro EDM can be controlled by intelligently choosing the machining parameters. The machining parameters include electrical parameters such as voltage, peak current, capacitance (for RC circuits), pulse ON time, pulse OFF time, pulse interval (for transistor-based circuits), and nonelectrical parameters like spindle speed, tool feed rate, dielectric flow rate and flushing type, and the type of dielectric. Performance of micro EDM can be assessed using MRR, tool wear ratio (TWR), surface roughness, recast layer, and geometrical distortions due to tool wear. Each of the machining parameters has a unique relationship with the performance parameters. Carefully selecting the optimum machining parameters helps to lower the machining time and increase the surface quality.

1.5 Micro EDM Variants

Even though we use the term micro EDM as a single machining technique, it is known to be a collection of different types of strategies with unique structure and capabilities. Micro EDM can be divided into the following categories.

1.5.1 Micro Die-Sinking EDM

In Micro Die-Sinking EDM, the tool electrode with a particular geometry is plunged to the workpiece without tool rotation. The shape of the tool will be printed as a mirror image on the workpiece with the help of electrical discharges (Maradia et al., 2012). This is considered to be one of the simplest and quickest forms of micro EDM without considering the tool fabrication time. However, the lack of tool rotation results in more number of arc discharges, and the tool is susceptible to welding with the workpiece surfaces at a higher voltage, capacitance, or current rating.

1.5.2 Micro EDM Milling

In Micro EDM Milling, tool electrodes of simple geometry (cylindrical rods) are used to trace out complex shapes by scanning over the workpiece surface along the programmed tool path. This eradicates the need for complex tool fabrication processes. The effect of tool wear is dominant in micro EDM milling, as the linear wear will result in depth difference and corner wear will result in shape difference. A tool wear compensation system has to be implemented to avoid the errors due to tool wear (Pham et al., 2004).

1.5.3 Microwire EDM

In Microwire EDM, a microwire is used to slice out the microfeatures. The continuously traveling wire reduces the effects of tool wear. Sharp corners can be made using microwire EDM. However, frequent wire rupture, wire vibrations, and the need of starting holes make it challenging to use in all machining requirements.

1.5.4 Micro Electrodischarge Grinding

In Micro Electrodischarge Grinding, the polarity of the tool is reversed to positive during machining. This process is mainly used for on-machine fabrication of microtool electrodes. There are four types of micro electrodischarge grinding (EDG) processes. Block EDGs (stationary and moving) (Asad et al., 2007), WEDG (Masuzawa et al., 1985), and disc EDG (Masuzawa et al., 2002; Pham et al., 2004). All the variants of micro EDM are briefly explained in Table 1.2 with applications and schematic representations. Table 1.3 shows the capabilities and machining characteristics of different micro EDM variants.

TABLE 1.2

Machining Variants of Micro EDM and Their Applications

Variant	Working	Applications	Schematic Representation
Micro die-sinking EDM	Microelectrode with desired features is plunged to the workpiece to produce corresponding mirror images	Microgears, microinjection molds, embossing molds, etc.	
Micromilling EDM	Rotating microelectrode with a simple geometry follows a predetermined tool path to produce complex microfeatures	Microfluidic channels, micropillars, other microfeatures, etc.	

(Continued)

TABLE 1.2 (*Continued*)

Machining Variants of Micro EDM and Their Applications

Variant	Working	Applications	Schematic Representation
Micro EDM drilling	Rotating microelectrode with simple geometry is plunged to the workpiece to produce shallow or deep microholes	Microholes for inkjet nozzles, reverse EDM tool electrodes, etc.	
Reverse micro EDM	Tool electrode with circular or noncircular microholes is plunged into producing micropillars	Fabrication of micropillars for biomedical applications. Fabrication of EDM or ECM tool electrodes for microdrilling of a series of holes, etc.	
Planetary micro EDM	Planetary motion is given to the microtool electrode to cut high-aspect-ratio microholes of a circular or noncircular geometry	Injection nozzles, microfluid systems, starting holes for μ-WEDM, etc.	

(*Continued*)

TABLE 1.2 (*Continued*)

Machining Variants of Micro EDM and Their Applications

Variant		Working	Applications	Schematic Representation
Micro wire EDM		Continuously moving microwire cuts microfeatures by traveling along a programmed tool path	Forming tools, stamping tools, spinning nozzles, etc.	
Micro EDG	Stationary block µ-EDG	Fabrication of micro electrodes using a stationary sacrificial block	On-machine fabrication of microelectrodes for micro EDM drilling and milling	
	Moving block µ-EDG	Fabrication of microelectrodes using a moving sacrificial block	On-machine fabrication of microelectrodes for micro EDM drilling and milling	

(*Continued*)

TABLE 1.2 (*Continued*)

Machining Variants of Micro EDM and Their Applications

Variant	Working	Applications	Schematic Representation
Wire micro EDG	Fabrication of microelectrodes using a continuously moving microwire	Cavities for microinjection molding. Embossing or coining tools, pin electrodes, rolling tools, etc.	
Disc micro EDG	Fabrication of microelectrodes using a rotating disc	On-machine fabrication of microelectrodes for micro EDM drilling and milling	

Source: Uhlmann et al. (2005).

TABLE 1.3

Different Micro EDM Variants and Their Capabilities

Micro EDM Variant	Min. Feature Size (μm)	Max. Aspect Ratio	Surface Quality R_a (μm)	Machining Capability
WEDM	30	~100	0.07–0.2	2½D Tapered surfaces with maximal angles of 150°
Die-sinking EDM	~20	~15	0.05–0.3	3D Free-form surfaces, undercut possible by planetary erosion, limited by electrode manufacturing
EDM milling	~20	10	0.2–1	3D free-form surfaces
EDM drilling	5	~25	0.05–0.3	Deep holes
WEDG	3	30	0.5	Axisymmetrical structures, minimum possible electrode diameter

Source: Liu et al. (2010) and Uhlmann et al. (2005).

1.6 Advantages

1. Low capital cost required compared to lithographic processes.
2. High flexibility in terms of product design.
3. No need to fabricate masks and special fixtures.
4. Free-form surfaces can be machined easily with the help of advanced CNC control systems.
5. Any conductive material can be machined regardless of the workpiece hardness.
6. Very small heat-affected zone.
7. On-machine tool fabrication is possible, which reduces tool-handling errors.
8. No burr formation.
9. No tool breakage due to tool–workpiece contact.
10. Thin walls can be machined without distortion.
11. High-aspect-ratio structures can be fabricated without tool breakage.

1.7 Challenges

1. Tool wear creates dimensional inaccuracies in the machined microfeature.

2. The workpiece must possess a certain level of electrical conductivity.

3. Formation of the recast layer will cause surface irregularities.

4. Surface integrity is affected by microcracks and micropores due to thermal stress.

5. The MRR is low compared to other machining techniques.

6. Debris accumulation in the machining zone has to be controlled to reduce the number of harmful discharges.

7. The spark gap has to be kept constant for stable machining.

1.8 Conclusion

In this chapter, an overview of the micromachining techniques is presented. The basic principle of EDM is discussed in detail, and an elaborate history of the machining process is detailed. Comparison of EDM and micro EDM techniques is presented. Discussion on general components of micro EDM will help to understand the technologies associated with the process. Principles and machining capabilities of different micro EDM variants are summarized. Finally, the advantages and challenges of using EDM for micromachining are discussed.

References

Ali, M.Y., Hung, N.P., 2000. Fabrication of micro cavities by micro electrical discharged machining. *In International Conference on Manufacturing*, Bangladesh.

Allen, D.M., Almond, H.J.A., Bhogal, J.S., Green, A.E., Logan, P.M., Huang, X.X., 1999. Typical metrology of micro-hole arrays made in stainless steel foils by two-stage micro-EDM. *CIRP Ann. Manuf. Technol.* 48, 127–130.

Amon, A., Alesch, F., 2017. Systems for deep brain stimulation: Review of technical features. *J. Neural Transm.* 124, 1083–1091.

Anders, A., 2003. Tracking down the origin of arc plasma science II}. Early continuous discharges. *IEEE Trans. Plasma Sci.* 31, 1060–1069.

Asad, A.B.M.A., Masaki, T., Rahman, M., Lim, H.S., Wong, Y.S., 2007. Tool-based micro-machining. *J. Mater. Process. Technol.* 192–193, 204–211.

Chen, W., Lam, R.H.W., Fu, J., 2012. Photolithographic surface micromachining of polydimethylsiloxane (PDMS). *Lab Chip* 12, 391–395.

Chow, H.M., Yan, B.H., Huang, F.Y., 1999. Micro slit machining using electro-discharge machining with a modified rotary disk electrode (RDE). *J. Mater. Process. Technol.* 91(1–3), 161–166.

Chung, D.K., Kim, B.H., Chu, C.N., 2007. Micro electrical discharge milling using deionized water as a dielectric fluid. *J. Micromech. Microeng.* 17, 867–874.

Egashira, K., Hosono, S., Takemoto, S., Masao, Y., 2011. Fabrication and cutting performance of cemented tungsten carbide micro-cutting tools. *Precis. Eng.* 35, 547–553.

Egashira, K., Taniguchi, T., Hanajima, S., Tsuchiya, H., Miyazaki, M., 2006. Planetary EDM of micro holes. *Int. J. Elec. Mach.* 11, 15–18.

Feynman, R., 1991. There's plenty of room at the bottom. *Science, 254,* 1300-1301.

Furutani, K., Mohri, N., Higuchi, T., 1997. Self-running type electrical discharge machine using impact drive mechanism. *Proceedings of the IEEE/ASME 1st International Conference Advanced Intelligent Mechatronics,* Tokyo, Japan, 88.

Ghosh, A., Mallik, A.K., 1985. *Manufacturing Science.* East-West Press Private Limited, New Delhi.

Gu, H., Duits, M.H.G., Mugele, F., 2011. Droplets formation and merging in two-phase flow microfluidics. *Int. J. Mol. Sci.* 12, 2572–2597.

Guo, Y., Ling, Z., Zhang, X., 2016. A novel PWM power amplifier of magnetic suspension spindle control system for micro EDM. *Int. J. Adv. Manuf. Technol.* 83, 961–973.

Han, F., Wachi, S., Kunieda, M., 2004. Improvement of machining characteristics of micro-EDM using transistor type isopulse generator and servo feed control. *Precis. Eng.* 28, 378–385.

Heckele, M., Schomburg, W.K., 2004. Review on micro molding of thermoplastic polymers. *J. Micromech. Microeng.* 14, R1–R14.

Hsu, T.-R., 2002. Miniaturization–a paradigm shift in advanced manufacturing and education. *International Conference on Advanced Manufacturing Technologies and Education in the 21st Century,* Chia-Yi, Taiwan.

Huo, D., 2013. *Micro-Cutting: Fundamentals and Applications.* John Wiley & Sons, Chichester.

Jahan, M.P., Rahman, M., Wong, Y.S., 2011. Study on the nano-powder-mixed sinking and milling micro-EDM of WC-Co. *Int. J. Adv. Manuf. Technol.* 53, 167–180.

Jahan, M.P., Rahman, M., Wong, Y.S., 2014. *Micro-Electrical Discharge Machining (Micro-EDM): Processes, Varieties, and Applications, Comprehensive Materials Processing.* Elsevier, Amsterdam.

Jahan, M.P., Wong, Y.S., Rahman, M., 2009. A study on the quality micro-hole machining of tungsten carbide by micro-EDM process using transistor and RC-type pulse generator. *J. Mater. Process. Technol.* 209, 1706–1716.

Jain, V.K., Sidpara, A., Balasubramaniam, R., Lodha, G.S., Dhamgaye, V.P., Shukla, R., 2014. Micromanufacturing: A review—part I. *Proc. Inst. Mech. Eng. Part B J. Eng. Manuf.* 228, 973–994.

Jameson, E.C., 2001. *Electrical Discharge Machining.* Society of Manufacturing Engineers, Dearborn, MI.

Kunieda, M., Hayasaka, A., Yang, X.D., Sano, S., Araie, I., 2007. Study on nano EDM using capacity coupled pulse generator. *CIRP Ann. Manuf. Technol.* 56, 213–216.

Kunieda, M., Lauwers, B., Rajurkar, K.P., Schumacher, B.M., 2005. Advancing EDM through fundamental insight into the process. *CIRP Ann. Manuf. Technol.* 54, 64–87.

Kuo, C.L., Huang, J.D., 2004. Fabrication of series-pattern micro-disk electrode and its application in machining micro-slit of less than 10 μm. *Int. J. Mach. Tools Manuf.* 44, 545–553.

Kurafuji, H., Masuzawa, T., 1968. Micro-EDM of cemented carbide alloys. *J. Jpn. Soc. Electr. Mach. Eng.* 3, 1–16.

Lin, J., Zhang, C., Yan, Z., Zhu, Y., Peng, Z., Hauge, R.H., Natelson, D., Tour, J.M., 2013. 3-Dimensional graphene carbon nanotube carpet-based microsupercapacitors with high electrochemical performance. *Nano Lett.* 13, 72–78.

Liu, K., Lauwers, B., Reynaerts, D., 2010. Process capabilities of micro-EDM and its applications. *Int. J. Adv. Manuf. Technol.* 47, 11–19.

Llanes, L., 2001. Influence of electrical discharge machining on the sliding contact response of cemented carbides. *Int. J. Refract. Met. Hard Mater.* 19, 35–40.

Manivannan, R., Pradeep Kumar, M., 2018. Improving the machining performance characteristics of the μEDM drilling process by the online cryogenic cooling approach. *Mater. Manuf. Process.* 33, 390–396.

Maradia, U., Boccadoro, M., Stirnimann, J., Beltrami, I., Kuster, F., Wegener, K., 2012. Die-sink EDM in meso-micro machining. *Procedia CIRP* 1, 166–171.

Masuzawa, T., 2000. State of the art of micromachining. *CIRP Ann. Technol.* 49, 473–488.

Masuzawa, T., Fujino, M., Kobayashi, K., Suzuki, T., Kinoshita, N., 1985. Wire electro-discharge grinding for micro-machining. *CIRP Ann. Manuf. Technol.* 34, 431–434.

Masuzawa, T., Okajima, K., Taguchi, T., Fujino, M., 2002. EDM-lathe for micromachining. *CIRP Ann. Manuf. Technol.* 51, 355–358.

Meeusen, W., Reynaerts, D., van Brussel, H., 2002. CAD tool for the design and manufacturing of freeform micro-EDM electrodes. *Symposium on Design, Test, Integration and Packaging of MEMS/MOEMS 2002*, Cannes-Mandelieu, France, 105–113.

Mhanna, R., Qiu, F., Zhang, L., Ding, Y., Sugihara, K., Zenobi-Wong, M., Nelson, B.J., 2014. Artificial bacterial flagella for remote-controlled targeted single-cell drug delivery. *Small* 10, 1953–1957.

Mishra, S., Yadava, V., 2015. Laser Beam MicroMachining (LBMM) - a review. *Opt. Lasers Eng.* 73, 89–122.

Morgan, C., Shreve, S., Vallance, R.R., 2003. Precision of micro shafts machined with wire electro-discharge grinding. *Proc. Winter Top. Meet. Mach. Process. Micro-Scale Meso-Scale Fabr. Metrol. Assem. Am. Soc. Precis. Eng.* 28, 26–31.

Ori, R.I., Itoigawa, F., Hayakawa, S., Nakamura, T., Tanaka, S.-I., 2004. Development of advanced alloying process using micro-EDM deposition process. *ASME 7th Biennial Conference on Engineering Systems Design and Analysis*, Manchester, American Society of Mechanical Engineers, 365–370.

Pham, D.T., Dimov, S.S., Bigot, S., Ivanov, A., Popov, K., 2004. Micro-EDM - recent developments and research issues. *J. Mater. Process. Technol.* 149, 50–57.

Pradhan, B.B., Bhattacharyya, B., 2008. Improvement in microhole machining accuracy by polarity changing technique for microelectrode discharge machining on Ti-6Al-4V. *Proc. Inst. Mech. Eng. Part B J. Eng. Manuf.* 222, 163–173.

Priestley, J., 1769. The history and present state of electricity, with original experiments. J. Dodsley, J. Johnson and J. Payne, and T. Cadell, London.

Qin, Y., 2010. *Micromanufacturing Engineering and Technology*. William Andrew, Oxford.

Raju, L., Hiremath, S.S., 2016. A state-of-the-art review on micro electro-discharge machining. *Procedia Technol.* 25, 1281–1288. doi:10.1016/j.protcy.2016.08.222.

Rajurkar, K.P., Levy, G., Malshe, A., Sundaram, M.M., McGeough, J., Hu, X., Resnick, R., DeSilva, A., 2006. Micro and nano machining by electro-physical and chemical processes. *CIRP Ann. Manuf. Technol.* 55, 643–666.

Rajurkar, K.P., Yu, Z.Y., 2000. 3D micro-EDM using CAD/CAM. *CIRP Ann. Manuf. Technol.* 49, 127–130.

Rasheed, M.S., 2013. Comparison of micro-holes produced by micro-EDM with laser machining. *Int. J. Sci. Mod. Eng.* 1, 14–18.

Razali, A.R., Qin, Y., 2013. A review on micro-manufacturing, micro-forming and their key issues. *Procedia Eng.* 53, 665–672.

Reynaerts, D., Van Brussel, H., 1997. Microstructuring of silicon by electro-discharge machining (EDM)—Part I: Theory. *Sens. Actuators A Phys.* 60, 212–218.

Schumacher, B.M., Krampitz, R., Kruth, J.P., 2013. Historical phases of EDM development driven by the dual influence of "Market Pull" and "Science Push." *Procedia CIRP* 6, 5–12.

Sen, M., Shan, H.S., 2005. A review of electrochemical macro- to micro-hole drilling processes. *Int. J. Mach. Tools Manuf.* 45, 137–152. doi:10.1016/j.ijmachtools.2004.08.005.

Sheu, D.Y., 2008. High-speed micro electrode tool fabrication by a twin-wire EDM system. *J. Micromech. Microeng.* 18, 105014.

Sheu, D.Y., Cheng, C.C., 2013. Assembling ball-ended styli for CMM's tactile probing heads on micro EDM. *Int. J. Adv. Manuf. Technol.* 65, 485–492.

Takahata, K., Shibaike, N., Guckel, H., 2000. High-aspect-ratio WC-Co microstructure produced by the combination of LIGA and micro-EDM. *Microsyst. Technol.* 6, 175–178.

Takezawa, H., Hamamatsu, H., Mohri, N., Saito, N., 2004. Development of micro-EDM-center with rapidly sharpened electrode. *J. Mater. Process. Technol.* 149, 112–116.

Tong, H., Cui, J., Li, Y., Wang, Y., 2007, January. CAD/CAM integration system of 3D micro EDM. *2007 First International Conference on Integration and Commercialization of Micro and Nanosystems*, Hainan, China, 1383–1387.

Tsai, Y.Y., Masuzawa, T., 2004. An index to evaluate the wear resistance of the electrode in micro-EDM. *J. Mater. Process. Technol.* 149, 304–309.

Uhlmann, E., Piltz, S., Doll, U., 2005. Machining of micro/miniature dies and moulds by electrical discharge machining - recent development. *J. Mater. Process. Technol.* 167, 488–493.

Vaezi, M., Seitz, H., Yang, S., 2013. A review on 3D micro-additive manufacturing technologies. *Int. J. Adv. Manuf. Technol.* 67, 1721–1754.

Vikram Singh, A., Sitti, M., 2016. Targeted drug delivery and imaging using mobile milli/microrobots: A promising future towards theranostic pharmaceutical design. *Curr. Pharm. Des.* 22, 1418–1428.

Volkert, C.A., Minor, A.M., Editors, G., 2007. Focused ion beam microscopy and micromachining. *Mater. Res. Soc. Bull.* 32, 389–399.

Williams, K.R., Gupta, K., Wasilik, M., Si, P., Si, P., 2003. Etch rates for micromachining processing II. *J. Microelectromech. Syst.* 5, 256–269.

Wollenberg, G., Schulze, H.P., Lauter, M., 2001. Process energy supply with pulses smaller than 200 ns and their thermal effects on micro-EDM 200 ns and their thermal effects on micro-EDM. *17th International Conference on Computer-aided Production Engineering (CAPE)*, Wuhan.

Yamazaki, M., Suzuki, T., Mori, N., Kunieda, M., 2004. EDM of micro-rods by self-drilled holes. *J. Mater. Process. Technol.* 149, 134–138.

Yan, C., Zou, R., Yu, Z., Li, J., Tsai, Y., 2018. Improving machining efficiency methods of micro EDM in cold plasma jet. *Procedia CIRP* 68, 547–552.

Yang, G.Z., Liu, F., Lin, H.B., 2012. Research on an embedded servo control system of micro-EDM. *Appl. Mech. Mater.* 120, 573–577.

Yang, J., Yang, F., Hua, H., Cao, Y., Li, C., Fang, B., 2018. A bipolar pulse power generator for micro-EDM. *Procedia CIRP* 68, 620–624.

Yu, Z.Y., Masuzawa, T., Fujino, M., 1998. Micro-EDM for three-dimensional cavities - development of uniform wear method. *CIRP Ann.* 47, 169–172.

Zhang, C., Xing, D., Li, Y., 2007. Micropumps, microvalves, and micromixers within PCR microfluidic chips: Advances and trends. *Biotechnol. Adv.* 25, 483–514.

Zhang, P., Ngai, T.L., Ding, Z., Li, Y., 2014. Erosion craters on Ti3SiC2anode. *Phys. Lett. Sect. A Gen. At. Solid State Phys.* 378, 2417–2422.

Zhao, W., Yang, Y., Wang, Z., Zhang, Y., 2004. A CAD/CAM system for micro-ED-milling of small 3D freeform cavity. *J. Mater. Process. Technol.* 149, 573–578.

2

General Components of Machine Tool

2.1 Introduction

Electro discharge machining (EDM) removes material by melting and vapor-ization with the action of uniformly distributed electrical discharges. To pro-duce these discharges, a DC power supply has to be provided between two electrically conductive electrodes submerged in dielectric fluid. To distribute the spark discharges uniformly over the tool surface, without transforming those into arc discharges, a pulse generator circuit has to be attached. To get the intended part shape, the tool electrode has to be moved along the designated tool path with greater accuracy. An accurate computer numeri-cal controlled (CNC) positioning system with adequate feedback control will serve that purpose. The dielectric fluid has to be pumped to the machining location, and the debris-contaminated fluid has to be filtered and recircu-lated using an efficient dielectric fluid delivery system. During EDM, a con-stant gap called spark gap has to be maintained between the tool and the workpiece, which demands a reliable servo control mechanism. The work-piece and the tools have to be constantly monitored with appropriate metrol-ogy systems to ensure dimensional accuracy of the machined product. Moreover, the machine tool vibration has to be damped before affecting the machining accuracy. The thermal expansion of the machine tool elements has to be kept to a minimum to avoid runout errors and tool positioning errors. Each of these requirements is fulfilled with dedicated components in an EDM machine tool, as shown in Figure 2.1. A detailed workflow of the micro EDM components is shown in Figure 2.2. Even though the basic mechanism of material removal is considered to be the same in both conven-tional EDM and micro EDM, the latter requires stringent control over the machining parameters and machining environment.

2.1.1 Discharge Energy/Discharge Frequency/Machining Parameters

To fabricate features with dimensions in the order of tens of microns, the vol-ume of the discharge crater has to be kept as small as possible. This can only be achieved by reducing the discharge energy to a minimum level. The pulse

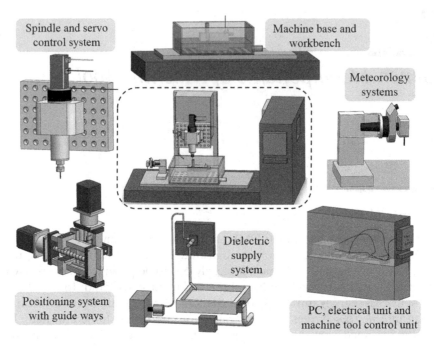

FIGURE 2.1
General components of micro EDM machine tool.

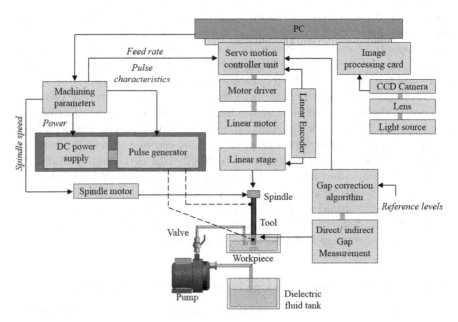

FIGURE 2.2
Detailed workflow of micro EDM components.

duration in micro EDM has to be as short as nanoseconds. So, any possible delay in the pulse generator circuits cannot be accepted. The pulse frequency has to be high enough to get sufficient material removal rate (MRR). The pulse energy also determines the amount of material eroded from the tool electrode (tool wear). Tool wear leads to dimensional deviations in the final microfeature (Kunieda et al., 2005). For precision machining, appropriate machining parameters have to be selected along with a suitable pulse generator circuit. EDM machine tool is controlled by the coordinated performance of different subsystems. CNC system generates the tool motion trajectory from the G and M codes written to specify the sequence of machining operations. The position control system takes input from the computer aided design and manufacturing (CAD/CAM) systems to position the tool precisely to the intended location. The discharge control system monitors and controls the pulse sequence, machining parameters, and the gap for smooth machining (Guo et al., 2006).

2.1.2 Precision in Movement

Since the tolerance range for component is very small, the tool/workpiece positioning system must have a resolution as high as possible (in the order of nanometers) to follow the exact tool path. A position feedback system with high resolution (<0.1 µm) has to be employed to avoid positioning errors. The machine tool must have high manufacturing accuracy (straightness and surface quality of the guideways and carriages). Possible backlash in the axis movements has to be prevented to avoid positioning errors (Zhang et al., 2015).

2.1.3 Spark Gap Control

The spark gap in micro EDM is very small when compared to that in conventional EDM. This demands a highly sensitive servo control system for measurement of the spark gap and to maintain the gap to the allowable range (Zhou and Han, 2009). The spark detection and response must be quick and accurate to avoid arcing, short circuits, and tool bending (Hashim et al., 2015).

2.1.4 Tool Wear Compensation

Tool wear compensation in micro EDM is essential to keep the dimensional and shape accuracy. As the size of the feature reduces, the errors due to tool wear become very serious. To avoid inaccuracies, an appropriate tool wear compensation mechanism (online or offline) has to be employed (Narasimhan et al., 2005).

2.1.5 Thermal Deformations

Thermal deformations of the machine tool elements have a larger effect on the machining accuracy as it causes runout errors and misalignment errors in spindle, carriages, guideways, and other machine tool elements.

Compared to conventional EDM, these errors will have serious implications in micro EDM accuracy, and it has to be avoided by selecting materials with low thermal expansion for machine structure. Sources of thermal errors have to be regularly monitored (Möhring et al., 2015), and temperature in the machining environment should be controlled to reduce the possible thermal gradients (Pham et al., 2004).

2.1.6 Other Considerations

The meteorology systems must have better magnification and resolution to measure the machine feature size and monitor the tool condition. The dielectric system must have the capability to flush out debris during machining of delicate and intrinsic parts. For long electrodes, electrode support subsystem (tool guide) has to be used to avoid eccentric rotation. To avoid misalignment and bending of tool electrode during fixing, on-machine tool processing system has to be employed. Moreover, to make an efficient ultraprecision EDM machine tool, the machine structure dynamics as well as safety, ergonomics, and quality control factors should be considered. Therefore, micro EDM machine tools demand more sophisticated components compared to conventional EDM machine tools. The structural configuration and building materials for the precision EDM machine tool are selected by analyzing the maximum travel distance of the axes, the dynamic interaction between the machine tool spindle, position control mechanisms and tool/workpiece fixtures. The research to improve the process capabilities of micro EDM is thus focused to resolve the key issues related to the subsystems. Some of the key issues regarding the subsystems of micro EDM are listed in Figure 2.3. Figure 2.4 shows the causes and effects of possible errors in micro EDM. The directions of research fields in the development of micro EDM subsystems are listed in Table 2.1.

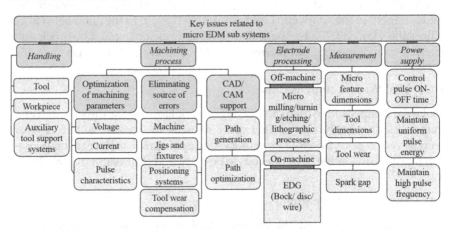

FIGURE 2.3
The main areas of concern in micro EDM research.

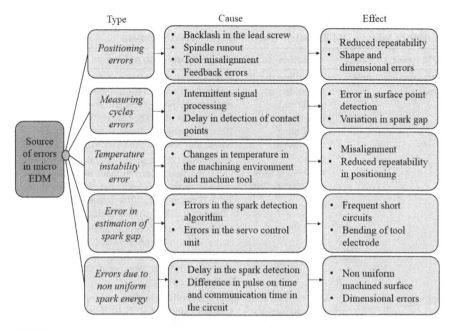

FIGURE 2.4
Sources and effects of possible errors in micro EDM.

TABLE 2.1

Outline of Micro EDM Research Towards Process Development

EDM Research Fields	Challenges	The Direction of Current Research
Surface quality and integrity	Avoiding microcracks on the workpiece material surface	Reducing the discharge intensity with the use of better pulse generators
		Reducing the crater size by developing power generators for very small discharge energy (Cheong et al., 2018; Chung et al., 2011; Giandomenico et al., 2018; Jahan et al., 2009a,c,d,e; Shah et al., 2007; Yang et al., 2018b)
	Increasing the machining accuracy and surface finish	The possibility of using additives in the dielectric fluid to increase the surface quality (Ali et al., 2011; Chow et al., 2008; Cyril et al., 2017; Modica et al., 2018; Mohammad, 2012; Prihandana et al., 2009a,b, 2011; Tan and Yeo, 2011)
		Optimizing the machining conditions for better surface finish, overcut, and tool wear (Ay et al., 2013; Azad and Puri, 2012; Feng et al., 2017; Manivannan and Kumar, 2016; Mehfuz and Ali, 2009; Natarajan and Suresh, 2015; Nirala et al., 2012; Singh et al., 2018; Tamang et al., 2017)

(Continued)

TABLE 2.1 (*Continued*)

Outline of Micro EDM Research Towards Process Development

EDM Research Fields	Challenges	The Direction of Current Research
MRR and machining time	Controlling the gap state for stable and uninterrupted machining operation to reduce machining time and increase the MRR	Development of better servo control systems and pulse discrimination systems to predict and control the gap state (Bissacco et al., 2010; Byiringiro et al., 2009; Mahardika and Mitsui, 2008; Mechanics, 2014; Moylan et al., 2005; Nirala and Saha, 2016; Yeo et al., 2009a)
		Optimization of the process parameters towards better machining performance regarding MRR and tool wear rate (Feng et al., 2017; Lin et al., 2015; Perveen et al., 2018; Tamang et al., 2017; Ubaid et al., 2018; Unune et al., 2018)
		Development of theoretical, analytical, and numerical models (Allen and Chen, 2007; Bigot et al., 2016; Guo et al., 2014; Klocke et al., 2018; Kurnia et al., 2008; Wang et al., 2009; Yeo et al., 2007a; Yu et al., 2003)
Tool wear	To reduce the tool wear to increase the dimensional accuracy and shape accuracy	Online and offline tool wear compensation systems (Aligiri et al., 2010; Bissacco et al., 2010; Heo et al., 2009; Jung et al., 2008; Mathai et al., 2018; Nguyen et al., 2013; Pham et al., 2007; Wang et al., 2017; Yan and Lin, 2011; Malayath et al., 2019)
		Development of insulated tools (Yuangang et al., 2009; Ferraris et al., 2013)
Combining different machining strategies	Utilizing the advantages of the different microfabrication processes to enhance micro EDM quality	Combining other micromanufacturing techniques (Al-Ahmari et al., 2016; Hao et al., 2008; Heinz et al., 2011; Kim et al., 2006b; Li et al., 2006; Liang et al., 2002)
New areas of applications	Extending the capability of micro EDM process towards different applications	Nanoparticle fabrication, fabrication of microstructures and microdevices, cutting tools, molds and punching tools, alloying of different materials, etc. (Chen et al., 2008; Chi et al., 2008; Hung et al., 2011; Jeong et al., 2009; Kumar et al., 2017; Lin et al., 2010; Peng et al., 2008; Sheu and Cheng, 2012; Tong et al., 2013)

2.2 Machine Structure, Vibration Dampers, and Spindle Unit

Machine structure is comprised of machine base, vibration isolation systems, positioning system support structure, machine support structure, spindle case, column, carriages, and workpiece table. Considering the functions of different parts in a machine structure, the elements can be grouped as follows (Kim et al., 2006a):

- Group 1: These are elements that serve as the foundation for different subsystems to be mounted on. These may include machine base, support structure for positioning systems, etc.
- Group 2: This includes machine elements that are confined in housings inside which various components are assembled to form a machine subsystem (e.g. spindle assembly).
- Group 3: These are machine tool elements used for fixing the workpiece and the tool.

The structure offers mechanical support to the machine tool elements. The structure elements and configuration for a machine tool are selected according to the material removal method and the amount of forces associated with it. The criteria for selecting the machine structure can be summarized by considering the following factors:

1. Stiffness and damping property of the structure material
2. Structural connectivity of the machine tool
3. Dynamics of the structural elements.

To have a sound performance, a machine structure must have high loop stiffness, high vibration damping capability, a closed-loop structural configuration with a symmetrical design, efficient elimination of heat concentration zones, stable performance of structural elements, and isolation from the influence of external factors including its surrounding environment. The main factors to be considered in selecting a machine structure for machine tools in micromachining can be summarized as follows (Möhring et al., 2015):

1. Structural rigidity: The structure must retain its shape and dimensions for a long period without any deflections and corrosion. The primary duty of the machine structure is to keep the rigidity of the geometric configuration under static, dynamic, and thermal loads. The machine frame should be able to absorb any undesirable disturbances that would otherwise propagate to the machining zone.
2. The interfaces or matting surfaces in the machine tool structure should be set up with high dimensional accuracy, straightness, and surface finish.
3. The configuration of the machine tool must have the endurance to vibrations as well as deflection due to self-weight, without exceeding the safe limit of working stresses. EDM machines popularly use the gantry structure to ensure better structural stability.

4. Good structural loop stiffness: The structural loop in a machine tool is comprised of spindle shaft, spindle bearing, housing, spindle fixture, guideway, support frame, motion drives, and tool/workpiece fixtures. The connectivity of the machine structure elements must be maintained. All elements must possess high stiffness to avoid deflections due to self-weight or external load. In EDM, the tool and workpiece never come into contact unless there is a short circuit due to the difference in servo feed and MRR. Thus, the chance of propagation of load to the machine structural elements from the cutting forces is negligible. However, rotation of the spindle, movement of the axes, dielectric flow, the weight of the subsystems, and auxiliary attachments will exert load on the structural elements.

5. Damping characteristics: Damping the possible vibrations without affecting the structural integrity of the machine tool is one of the main concerns in machine tool manufacturing. This can be tackled by selecting appropriate materials for machine structure and employing separate dampers for isolating the vibrations. Cavities in the machine structure are often filled with oil and lead shots, for enhancing viscous damping properties. Shear plates and tuned mass dampers can be used to isolate vibrations without propagating it to the structural elements.

6. Minimizing thermal gradient: The machine structure must be able to reduce the thermal gradients in the system by eliminating areas of heat concentration and must help to attain thermal equilibrium as fast as possible.

7. Isolation from external environmental factors: The machining environment has to be kept free from external vibrations, loads, temperature variations, and dust particles.

2.2.1 Functions of Structural Elements

Machine base helps to damp the vibrations originating from moving parts, rotating spindle, or any external factors. Micro EDM has the major advantage of force-free machining (no contact between tool and workpiece). So, there is no chance of structural loading due to cutting forces during machining. This also removes the chance of chatter and mechanical wear of cutting tool edges, which are common in conventional machining. The force-free nature also becomes advantageous in the case of fixture design for the workpiece. Simple fixture design can be used for the workpiece.

2.2.2 Materials for Machine Tool Structure

Selection of materials for machine structure has a prominent role in determining the stability and soundness of the machine tool configuration.

Cast iron and granite are the popular materials used to make machine base because of their high stiffness. However, they are very heavy, and the machine tool weight also increases substantially. Reduction of the total weight of the machine tool is also considered a major parameter in selecting the machine structure material. However, this should be done without compromising the effective stiffness of the machine tool. The thumb rule is to make machine tools with minimum possible weight and adopt a configuration so that the weight is distributed in the machine tool to ensure uniform loading (Möhring et al., 2015). Lightweight polymer composites can be an alternative for cast iron or granite. It also possesses better damping characteristics and structural rigidity. To reduce the thermal expansion in high temperatures, materials such as super invar, synthetic granite (Chen et al., 2016), ceramics, Zerodur, and composite foam resin concrete (Kim et al., 2006a) are used in the manufacture of structural elements of ultraprecision machine tools.

Material for the machine tool base is selected by considering the temporal stability, homogeneity, stiffness, machinability, and manufacturing cost to produce it with a high degree of dimensional accuracy and surface quality (Huo, 2013). Stability in high temperature variations can be achieved by selecting the material with low thermal expansion coefficient and low specific heat capacity. Materials such as metal, stone, ceramic, polymer concrete, porous, and reinforced composite materials are commonly used. Among the materials used for machine structural elements, steel, cast iron, and other metals are popular choices. Even though this monopoly has been challenged by the emergence of new materials, the materials for mechanical interfaces, guides, joints, and bearings are still made of metals. Among these metals, welded steel and cast iron are considered to be the competitors. Granite is used in the manufacture of machine base because of its high damping characteristics. Granite frames also have low thermal expansion (0.006–0.008 mm/mK), and they are much more stable because of the absence of residual stress (Möhring et al., 2015). However, resistance to tensile loads and bending loads is a basic issue in machine frames made of natural stones. Mineral casting has lower thermal expansion compared to steel and cast iron, which makes it more stable in high thermal loads. Metal foams and fiber-reinforced plastics are the other alternatives to fabricate machine structures for precision machine tools. Details of these materials are summarized in Table 2.2 (Möhring et al., 2015; Huo, 2013).

Even though the micro EDM machine tools provide the advantage of chatter-free, cutting force-free machining environment, any possible vibration in the machine tool affects the machining accuracy significantly. Literature discussing the machine structural configuration of micro EDM and the effect of materials on the accuracy of machining is very rare. Kim et al. (2006a) used polymer-based, fiber-reinforced composites to fabricate micro EDM structural elements and discussed their advantages. It showed high specific stiffness and damping characteristics. Sandwich structures for

TABLE 2.2

Machine Tool Structure Materials and Their Areas of Application (Mohring et al., 2015; Huo, 2013)

	Gray Cast Iron	Steel	Mineral Casting	Granite	Carbon Fiber-Reinforced Plastic
Applications	Bed, column, slide, table, spindle casing	Bed, column, slide, table	Bed, column,	Bed, column	Slide, table, spindle casing
Compression strength (N/mm²)	600–1,000	250–1,200	140–170	70–300	–
Tensile strength (N/mm²)	150–400	400–1,600	25–40	30–35	400–2,400
Thermal coefficient of expansion (μm/mK)	10	12	12–20	6.5–8.5	–
Density (g/cm³)	7.15	7.85	2.1–2.4	2.9–3.0	1.6
Damping (logarithmic decrement)	0.003	0.002	0.02–0.03	0.0015	–
Heat conductivity (W/mK)	50	50	1.3–2.0	1.7–2.4	1–50
Modulus of elasticity (kN/mm²)	80–120	210	30–40	35–90	48–360

the machine structures are proposed, and the design is optimized using an experimental investigation consisting of several variables including varying composite geometries, stacking sequence, rib thickness, and rib geometry. An L-shaped joint is fabricated for combining the bed and column of the micro EDM machine tool, and vibration tests are conducted to analyze the performance of the machine structure.

2.2.3 Vibration Isolators/Dampers

Vibration isolators are used to avoid the propagation of external shock from the floor to the machine tool elements. Popular vibration damping methods include coil spring mounts, elastomeric isolation elements, and vibration mounts (www.vibrodynamics.com). The main applications of the vibration dampers include enhancing the effective machine stiffness to avoid errors due to variation in loads, minimizing the gradient in weight distribution, and providing stability to the machine tool structure. Elastomeric cushion pads are widely used to isolate external shocks from the machine tool

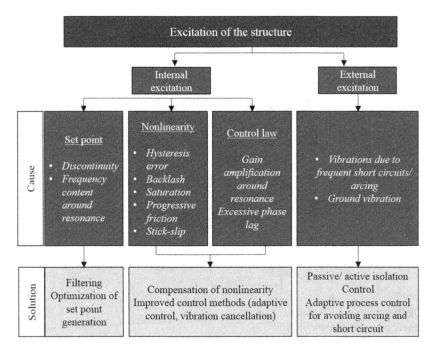

FIGURE 2.5
Causes of vibration and the remedies in a micro EDM machine tool (Altintas et al., 2011).

(www.vibrodynamics.com). Sometimes, the vibration in the machine structure gets amplified because of structural resonance and adversely affects the structural integrity of the machine tool. To avoid this, active or passive dampers have to be attached to the machine tool to increase the internal damping characteristics. Tuned mass dampers are a kind of devices used to reduce vibration in the machine tool. Figure 2.5 shows the sources of vibration in a micro EDM machine tool and the methods to reduce their influence.

2.2.4 Machine Spindle

The spindle assembly consists of a driving motor (integrated to the spindle shaft or belt driven from a separately placed motor), spindle housing, spindle bearings, spindle shaft, and tool clamping system (www.dynospindles.com/vault/technical/Book-of-Spindles). The spindle motor drives the spindle shaft, and the bearing system ensures smooth rotation at all ranges of speed. In micro EDM, an integrated AC servomotor is commonly used to drive the shaft. The tool clamping system can be either mechanical or pneumatic. The mechanical tool clamping system uses a collet-and-nut arrangement to hold the tool. The pneumatic variant makes use of compressed air-driven clamper to keep the tool in position. Spindle bearing is considered to be the most critical component in the spindle assembly. Cylindrical roller, tapered roller, and angular

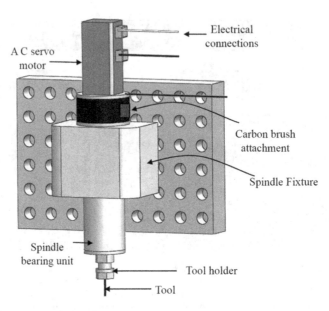

FIGURE 2.6
Schematic diagram of a spindle unit.

contact bearings are commonly used in precision machine tool spindles. High-speed spindles for precision microtools are often equipped with angular contact bearings. They can bear axial and radial loads. Tapered roller bearings are used where high load capacity and high stiffness of the bearing unit are essential). The primary duty of spindle housing is to locate the bearings accurately. Sometimes, the spindle housing has channels to facilitate the flow of coolants (water/oil/air) which takes away the heat generated by the spindle motor and other moving elements. Figure 2.6 shows the parts of a spindle unit.

Compared to other precision machine tools, EDM has some advantages and disadvantages regarding the spindle design. In EDM, the cutting forces are absent, which reduces the axial or radial loading on the spindle unit. However, for EDM, the tool has to be connected to the power supply. For this, special arrangements have to be made inside the rotating spindle for electrical connections. To keep the spindle motor and other machine elements safe, electrical isolation has to be provided inside the spindle unit. Frequent short circuits can result from tool–workpiece contact, and the shock generated can propagate to the spindle. Thus, special care needs to be taken to reduce the effect of shocks generated due to short circuits. Conventionally, this is achieved by a servo control unit (discussed in the next section) which retracts the tool electrode as soon as the circuit detects a short circuit. However, the delay in identifying the short circuit signals from the discharge characteristics will cause serious damage to the tool and workpiece. One of the alternatives is to develop an intelligent spindle unit in which the short-circuit detection can be directly coupled with the spindle unit. Guo and Ling (2016) developed a magnetic suspension system dedicated to micro EDM consisting of a pair of radial and thrust electromagnetic bearings with a magnetic couple linker assembled to the conventional spindle unit. In this, the radial bearings levitate the shaft in the air avoiding mechanical contact with the spindle case. Thrust bearing controls the tool motion. If the thrust-bearing current is positive, the spindle is fed downwards. During short circuits, the resultant axial force causes an upward tool movement. A pulse width modulation (PWM) wave is used to control the thrust currents to pull back the tool electrode (Guo et al., 2018).

2.2.4.1 Thermal Errors in the Machine Tool Spindle

Machine tool spindle experiences internal temperature variations due to heat generation by friction in the bearing unit as well as spindle motor, or external temperature variations due to environmental changes. There are different strategies to reduce the effect of external or internal thermal loads (Li et al., 2015),

1. Replace the metal ball bearings with ceramic bearings to reduce the effective friction and heat generation in the bearing unit.
2. Control the temperature gradient in the spindle unit by placing cooling jackets or cooling circuits around the spindle-bearing assembly.
3. Compensate for the possible thermal errors in the machine spindle by adjusting the tool–workpiece positions. This includes the steps of analyzing the temperature gradient in the spindle assembly, modeling the thermal error to find the relationship between temperature variations and thermal errors, predicting the axis deviations due to temperature changes, and incorporating the error compensation by altering the origin coordinates.

   Current research in machine tool spindles is focused on the development of intelligent spindle systems that can identify, learn, and control possible errors and continuously monitor the spindle health (Cao et al., 2017). Further development in this research will help to attain more stable micro EDM system with less short circuits, high machining accuracy, and higher machining rate. A general representation of intelligent spindle systems for machine tools is given in Figure 2.7.

2.3 Position Control and Servo Control System

The precision of a machine tool depends on the positioning of the cutting tool/spark/energy beam responsible for material removal. One of the revolutionary discoveries that led to the development of precision machine tools is the invention of computer numerical control (CNC) system. In the CNC

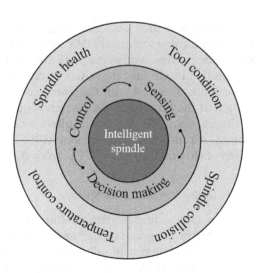

FIGURE 2.7
Representation of intelligent spindle unit system (Cao et al., 2017).

system, the computer associated with the machine tool reads, interprets, and executes a specific operation using a program written in alphanumerical codes. This includes the movement control of workpiece/tool and sending ON–OFF signals to various machine tool subsystems (e.g. coolant fluid delivery system). The CNC system must be capable of moving the X, Y, Z and rotary stages with high resolution and minimum inertial, hysteresis, and backlash errors. Precision movements in machine tools are achieved by using a servomechanism for motion control. The most popular servo control system in machine tools consists of a servomotor, a lead screw, a precise guideway, and a linear encoder for position feedback. Rotary motion of the motor is converted to linear motion with the help of the lead screw and nut mechanism. To reduce the errors due to backlash in the lead screw mechanism, a ball screw arrangement is often used in the positioning system. Precise guideways will ensure the smooth motion of the stage without any sideway tilting. The optical encoder will give the feedback to the servo controller unit to correct the position of the stage.

To handle the errors due to backlash, thermal load, and other factors, lead screw compensation has to be applied (Zhang et al., 2015). Even though this mechanism offers adequate results during machining, EDM demands more sophisticated servo control methods due to special characteristics of the machining method. Apart from positioning the tool electrode into the exact location with greater accuracy, the gap between the tool electrode and the workpiece (spark gap) has to be maintained to continue the material removal process smoothly. The difference in feed rate and MRR will result in short circuits, arcing, and tool bending and breakage. To avoid this, the spark gap should be monitored continuously, and the servo controller must be able to push forward or retract the tool electrode to keep the spark gap in the predefined levels. Moreover, in micro EDM, this gap is very small, which makes direct monitoring and control of spark gap more difficult. Being a stochastic process, the servo control demands more complex algorithms to control the spark gap.

The main components in the positioning system of micro EDM are as follows:

- Guideways
- Drive systems
- Feed motors and sensors
- Feed control systems.

2.3.1 Guideways

Guideways ensure smooth sliding of the stage and straightness of the movement. Guideways restrict sideway movements and damp the vibrations during the movement. The guideways of micro EDM must have the following characteristics (Altintas et al., 2011):

1. High dimensional accuracy and surface finish facilitating smooth sliding of the stages.
2. High stiffness to reduce deformation due to inertial forces.
3. Low coefficient of friction and ability to maintain corrosion-free, high wear-resistant surfaces for a long lifetime.
4. High toughness to absorb shocks due to unexpected collisions, short circuits, etc.

The popular types of guideways are detailed in the following (Altintas et al., 2011).

2.3.1.1 Frictional Guides

Frictional guides offer high load capacity, ability to withstand impact loads, and good damping characteristics. Uniform marks and deep lubrication slots on the sliding surface ensure uniform contact area and smooth movements.

2.3.1.2 Rolling Guides

Rolling guides with stationary or recirculating rollers are commonly used in precision machine tools. Recirculating rollers are preferred where the stroke length is comparatively larger. Rolling guides have the advantages of low friction, great load-bearing capacity, and stiffness, but they lack the damping ability.

2.3.1.3 Hydrostatic/Aerostatic Guides

In this type of guides, an oil/air layer is present between the sliding surfaces, which reduces the frictional coefficient and the chance of stick-slip.
Hydrostatic guides have better damping characteristics than the roller guides
in the direction normal to the sliding surface. This guiding system is sensitive
to temperature, as the viscosity of the oil changes with regard to the temperature variations in the machine tool elements. Thus, the vibration damping
characteristics and load-bearing capacity of these guides also depend on the
thermal loads in the system. Compared to rail guides, hydrostatic guides are
expensive and complex. In aerostatic guides, the stages are supported by a
thin film of air which provides very low friction during sliding. Due to their
low load-carrying capacity, aerostatic guides are not preferred in conventional
machine tools. As micro EDM often deals with very small or no cutting forces,
aerostatic guides has become a feasible choice. Vacuum preloading of air bearings ensures compact guideways with good vertical and angular stiffness.

2.3.1.4 Magnetic Levitation Guides

Here, the sliding elements are levitated with the help of a magnetic field
which removes the chance of contact, friction, and need of lubrication.
Magnetic levitation guides are often coupled with linear motor drives. The
control system for positioning system with magnetic levitation is very complex, as it demands 5 degrees of freedom control (Zhang et al., 2007).

2.3.1.5 Compliant Mechanisms

Compliant mechanisms are used in precision machine tools that demand
small stroke length. In these, the elastic properties of the elements determine
the effectiveness of the guiding system (Hongzhe et al., 2015). They have low
friction, less backlash and less slip-stick errors, but they lack load-bearing
capacity and have a small stroke length. Figure 2.8 shows a compliant mechanism used in micro EDM.

2.3.2 Drive Systems

Commonly, two kinds of machine tool drives are used in precision micro
EDM tools:

1. Lead screw mechanism with recirculating balls
2. Linear drives.

2.3.2.1 Ball Screw Mechanism

In a ball screw mechanism, the lead screw is supported by thrust bearings at
ends and a special nut with recirculating balls. One end of the lead screw will

FIGURE 2.8
A compliant mechanism developed by AGIE® for tool positioning in micro EDM (Altintas et al., 2011).

be connected to the drive motor. Backlash errors are avoided by preloading the nuts. Dimensional accuracy of the lead screw and nut is very important as the pitch errors will reflect as positioning error during machining. The main characteristics of ball screw drives are as follows (Zhang et al., 2015).

2.3.2.1.1 Advantages

- Low wear
- High mechanical efficiency
- Low heating
- Greater service life without stick-slip effect
- High reliability
- Less starting torque required
- Low coefficient of friction compared to sliding-type screws
- Backlash can be easily eliminated by compensation methods and preloading
- Provide longer thread life
- Suitable for applications with long cycle times like EDM.

2.3.2.1.2 Disadvantages

- Susceptible to vibrations
- Periodic maintenance is required
- Presence of foreign particles may deteriorate the thread life
- Requires a high level of lubrication.

2.3.2.2 Linear Motor

In micro EDM, as the tool and workpiece maintain a gap between them, the load-bearing capacity of the drive system does not have to be very high. Different technologies can be used as alternatives to the servomotors to make the system less complex, more responsive, and more reliable. A linear motor is capable of producing direct translatory motion. There are several kinds of linear motors available including Lorentz-type actuator, linear induction motors, and piezoelectric actuators. The same guide system used for ball screw mechanism can be applied while employing linear motors, and the screw, nut, coupling, thrust bearings, and motor shaft can be eliminated (Altintas et al., 2011). The absence of conversion systems reduces the chance of backlash errors and friction, but increases the reliability and lifetime of the positioning system. Very high sensitivity and quick response make the linear motors reliable positioning systems for micro EDM.

Hu et al. (2001) developed a linear electrostrictive servomotor for positioning and gap control in micro EDM, as shown in Figure 2.9a. Using the linear motor system, microholes are successfully machined on the sharp end of a stainless steel needle, as shown in Figure 2.9b. Hsieh et al. (2007) developed a dual linear motor drive for die-sinking EDM to improve the overall thrust and structural stiffness of machine tool drive. An adaptive control strategy is developed for linear motors for spark gap control (Hsue and Chung, 2009). A programmable logic device is used to discriminate the pulses, and effective pulse counting and servo voltage are used to control the jump height during tool retraction. Synchronous control of a twin parallel-axis linear servomechanism is elaborately discussed by Yao et al. (2011).

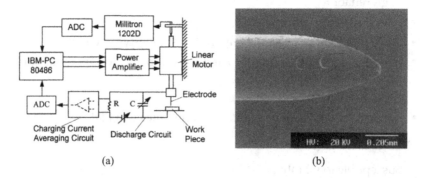

(a)   (b)

FIGURE 2.9
(a) Gap control circuit in a linear-motor-equipped micro EDM machine tool and (b) microholes drilled on a stainless steel needle end using a linear-motor-controlled micro EDM (Hu et al., 2001).

2.3.2.3 Local Actuators

To improve the response characteristics of tool positioning and retraction system in EDM, local actuators are recommended by various researchers. Masuzawa (1975) studied the possibility of using a voice coil actuator for EDM tool feed control. An impact drive mechanism and a direct drive mechanism are introduced for EDM feed control (Higuchi et al., 1991). Imai et al. (1996) used a high-frequency response piezoelectric actuator in EDM and reported that the machining rate is improved 1.5–2.5 times by high response jumping action of the tool electrode. Various micromotion technologies are invented for precise control of the ultraprecision machine tools including shape memory alloy actuators (Sadeghzadeh et al., 2012), ultrasonic motors (Maas, 2000), stick-slip actuators (Zhang et al., 2006), and complaint mechanisms (Hongzhe et al., 2015). However, most of the researchers employed piezoelectric actuators and maglev actuators for microfeed control in micro EDM. Imai et al. (2004) used a local actuator based on electromagnets with thrust coils for drilling microholes and fabricating micro dies. A microfeed technology based on inchworm mechanism is developed by Li et al. (2002).

2.3.2.4 Piezoelectric Actuators

Piezoelectric actuators can produce translatory motion with the help of shape-changing characteristics of piezoelectric materials in the electric field. The modified piezoactuator is used for micro EDM by Muralidhara et al. (2009). An electromechanical modeling is used to reduce the effect of hysteresis behavior inherent to piezoelectric systems. Two piezostacks are serially arranged, and a flexural link is used to amplify displacement up to 445 µm. The piezoelectric actuator is also used for micro electrodischarge grinding (EDG) control (Venugopal et al., 2016). Generally, local actuator is used as an additional feed drive in micro EDM systems.

Feed control strategy: The piezoelectric actuator serves as an additional capacitor in the circuit. When the supply is on, the actuator will also charge along with the capacitor. The piezoelectric crystals get elongated in proportion to the voltage. At optimum spark gap, the capacitor and the actuator are instantaneously discharged, and the actuator retracts to the original state. The voltage and current parameters are controlled using a self-adaptive system to maintain optimum elongation length and spark gap.

2.3.2.5 Maglev Actuators

The maglev actuators consist of upper and lower magnetic radial bearings, thrust magnetic bearing, electromagnet, permanent magnet, air motor, and displacement sensors, as shown in Figure 2.10 (Zhang et al., 2007).

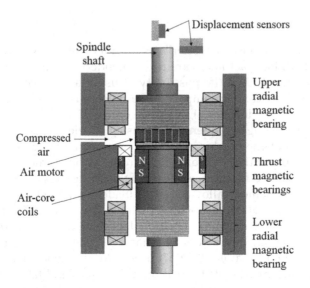

FIGURE 2.10
Schematic diagram of a Maglev actuator.

Using thrust and radial magnetic bearings, the spindle shaft can be placed without any mechanical contact with other elements. Two air coils and a permanent magnet are used to control feed movements, including tool retraction in micro EDM. The Lorentz force generated between the stator coils and the permanent magnets provides the driving force (Zhang et al., 2007). Zhang et al. (2008) used the maglev actuator for machining in the radial direction, thrust direction, and with planetary movements in high-speed EDM. Better control systems are then developed for precise control of the actuator for gap state control in EDM (He et al., 2010; Zhang et al., 2008).

2.3.2.6 Hybrid Systems

To combine the advantages of motors and actuators, hybrid positioning systems are developed. Hebbar et al. (2002) combined a brushless DC motor drive with a piezoelectric actuator. As an alternative to this technology, a hybrid system is developed with a conventional ball screw drive system for the Z-axis stage (for macrovertical feed) and a piezoelectric microvertical feed arrangement for worktable, as shown in Figure 2.11 (Fu et al., 2013, 2016; Herzig et al., 2017). The motor drive helps to position the actuator, and the actuator helps to provide a faster response towards servo control. A hybrid system comprising a linear motor and a piezoelectric actuator is used in servo scanning micro EDM (Tong et al., 2008). As shown in Figure 2.12, the local actuator improves the response characteristics of the motion control system and does effective real-time gap control. Linear motor helps to increase the travel distance and acceleration of the vertical stage.

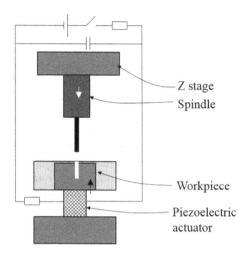

FIGURE 2.11
A hybrid position control system in which the actuator is connected to the workpiece table for a quick response.

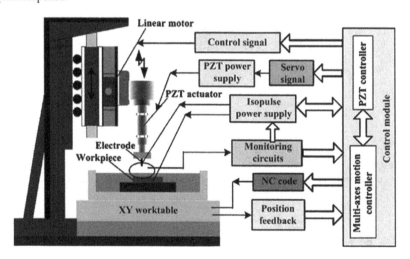

FIGURE 2.12
The configuration of a hybrid system with servomotor-controlled vertical stage and a piezoelectric quick response actuator used for servo scanning 3D micro EDM (Tong et al., 2008).

2.3.3 Sensors

To position the machining stages precisely, a position feedback system is essential. Optical encoders, laser interferometers, and syncroresolvers are commonly used to monitor the position of the stage (Altintas et al., 2011). The displacement information can be directly transmitted to the digital readout. As shown in Figure 2.13, the optical encoder system contains a metal/glass grating, an LED light source, and a photodetector. The grating will transmit or reflect light which is captured and evaluated by the photosensor. The encoder scale contains equally spaced reflective gratings. Among the two kinds of encoders, absolute encoders are the one with binarily coded gratings to find the position. In relative or incremental coders, a reference mark helps to find the current position by counting the equally spaced marks on the scale. The alignment of the reflected light with respect to the structured detector will give the position and direction of motion. Micro EDM machine tools demand a high-resolution encoder system to ensure precision motion. Interpolation of the sinusoidal pattern of the encoder readings will help to increase the resolution of optical encoders. However, the exposed linear encoders are vulnerable to contamination from oil, dust, and foreign particles. This will cause positional errors during machining. Appropriate error compensations and sealed and clean environment will reduce the effect of errors. The effect of contamination in the accuracy of the grating is shown in Figure 2.14.

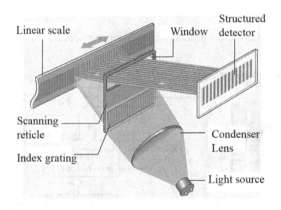

FIGURE 2.13
Working of a linear encoder system (www.industrial.embedded-computing.com).

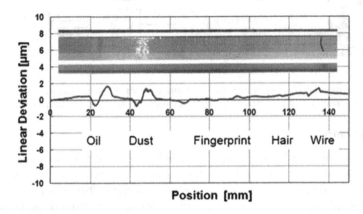

FIGURE 2.14
Optical encoder errors (www.heidenhain.us).

Feed drive velocity can be regulated by direct or indirect means. Rotary tachogenerators give voltage signals concerning the angular speed of the drive motor (Altintas et al., 2011). Indirect methods include differentiating the position measurement from optical encoders to calculate feed velocity. The source of errors in velocity estimation includes errors associated with encoder readings, quantization errors, and the presence of external harmonic error frequencies. To control structural vibrations, estimation of acceleration of the feed motion is necessary. Piezoceramic-based accelerometers are used to provide acceleration feedback signals (Altintas et al., 2011). Electromagnetic sensors are also used to estimate the relative acceleration between the table and the base.

To discriminate the discharge state in the spark gap, discharge current has to be monitored continuously. Serially connected shunt resistors, inductive transformers, and hall sensors are commonly used to estimate the change in current amplitudes due to machining disturbances (Altintas et al., 2011). Indirect measurement of the gap voltage can be realized by fixing a ring electrode coaxial to the tool electrode without any mechanical contact (Kunieda et al., 2007). Change in electrode potential will reflect in the potential measured from the ring electrode, as the two are connected through capacitance coupling. The stray capacitance produced by the coupled capacitors and voltage measuring probe is adjusted appropriately to get a proportional relationship between the gap voltage and the measured voltage. To reduce the noise in the acquired signal due to electrical discharges, a Kalman filter can be used (Xi et al., 2017).

2.4 Pulse Generators

To produce uniform discharges at different locations on the workpiece surface, a pulsed power supply is used in EDM. There are different methods to produce pulsed power supply. Rotary impulse pulse generators (Dunn and Clayton, 1964) produce sinusoidal voltage waveform using the principle of DC generators. However, the difficulty in controlling the wave characteristics makes it inappropriate for EDM. Relaxation circuits or RC circuits are capable of generating pulses in the form of a sawtooth with the help of a resistor–capacitor combination (Jahan et al., 2009b). The charging time of the capacitor is considered as the pulse OFF time and discharging time as the pulse ON time. A transistor-type pulse generator uses a transistor circuit to generate rectangular pulses with variable pulse durations (Jahan et al., 2009b). As the relaxation circuit and the transistor circuit have some inherent limitations, researchers developed hybrid circuits for better performance of the pulse generator system (Kunieda et al., 2007).

2.4.1 Transistor-Based Pulse Generator

In this type of generator, a field effect transistor (FET) is connected to some resistors and a power source as shown in Figure 2.15. The transistor circuit generates rectangular pulses, and the ON–OFF durations are controlled by FET. However, there exists a delay in transferring the information regarding the gap state due to the time constant in circuits. The discharge current can be controlled by increasing the number of transistors connected in the circuit (Jahan et al., 2014). It has the capability of producing high-frequency pulses, which increases the machining rate. The pulse characteristics can be easily varied using FET, which makes it more flexible to operate. However, the transmission delay sometimes exceeds the discharging time duration and results in nonuniform pulses (Han et al., 2004). Transistor-based pulse generator for wire electrodischarge grinding (WEDG) with two lines of FETs in the circuit is developed by Hara and Nishioki (2002). One of the FETs is used for detecting short-circuit current and the other for processing. Jahan et al. (2008) conducted a comparative study of an RC and transistor-type pulse generator by machining microholes on a tungsten carbide workpiece. Gap voltage, peak current, and duty ratio are the factors that affect machining performance. It is found that while reducing the voltage below 60 V, the machining micro EDM process becomes unstable and fails to erode material from workpiece surface even if the resistance is set to the lowest value (Jahan et al., 2008). In transistor circuits, the duty ratio of 0.5 produces stable pulses with uniform intensity because of adequate time for deionization and debris removal in the dielectric medium.

2.4.1.1 Important Characteristics of Transistor-Based Pulse Generators

- Rectangular pulses
- Time delay for capacitor charging is eliminated—precise control over pulse interval time
- High discharge frequency

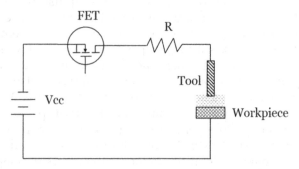

FIGURE 2.15
Transistor-based pulse generator circuit.

- High MRR.
- Pulse duration and discharge current can be controlled.
- The reverse flow of current or polarity shifting is restricted by the immediate cutoff of current by FET.
- Difficult to keep uniform pulses due to transmission delay in the circuit.

2.4.1.2 Pulse Characteristics

Figure 2.16 shows a representation of the ideal and realistic voltage and current pulses during EDM. The terminologies related to the pulse characteristics are explained as follows:

- Breakdown voltage: The voltage at which dielectric breakdown happens.
- Gap voltage: Voltage measured between the tool and the workpiece.
- Peak current: Maximum amplitude of current in the gap during a discharge period.
- Pulse ON time: The period when discharge occurs.

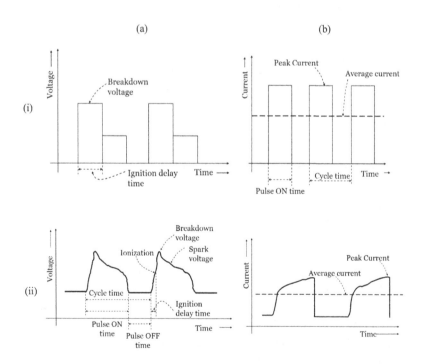

FIGURE 2.16
(i) Ideal and (ii) realistic pulse waveforms of (a) voltage and (b) current signal in transistor-based EDM circuits.

- Pulse OFF time: The period when discharge disappears and deionization of dielectric fluid starts.
- Ignition delay: It is the delay in time between start of a pulse and start of a discharge.

2.4.2 RC Pulse Generator

In a relaxation circuit, the DC voltage source is connected to a storage element (RC, LC, and RLC). The gap between tool and workpiece separated by a dielectric medium behaves as a switch. As the dielectric breakdown happens, the current begins to flow through the gap till the storage element is discharged. In an RC circuit, the pulse duration depends on the capacitor rating and the inductance of the wire (Rajurkar et al., 2006). The number of pulses per unit time is determined by charging and discharging times of the capacitor. To charge the capacitor quickly, the resistance can be reduced. It will lead to higher frequency of discharge pulses and eventually result in higher machining rate. However, reducing the resistance value below a critical value will lead to frequent arcing and destroys the stability of machining (Ghosh and Mallik, 1986). The discharge energy depends on the capacitance rating. In addition to the capacitors connected in the circuit, the tool holder–tool, the tool–workpiece, and the gap voltage measuring probe will add some stray capacitance. As the total charge–discharge depends on the equivalent capacitance, the stray capacitance will decide the minimum discharge energy possible in RC-type micro EDM machines (Rajurkar et al., 2006). Figure 2.17a shows the RC circuit diagram of an EDM pulse generator. Figure 2.17b shows the charging phase, and Figure 2.17c shows the discharging phase.

2.4.2.1 Important Characteristics of RC-Type Pulse Generators

Figure 2.18 shows the ideal and realistic representation of current and voltage pulses in RC circuits. The main characteristics of an RC-type pulse generator are as follows (Han et al., 2004):

- Short time required for the discharge of capacitor makes the RC circuit capable of producing very small discharge durations (in the order of nanoseconds).
- Discharge energy can be very small, as it depends on the capacitance rating and stray capacitance value.
- For finishing, stray capacitance can be used to generate pulses with ultrasmall discharge energy.
- Discharge frequency is low due to the charging time of the capacitor. This leads to low MRR.

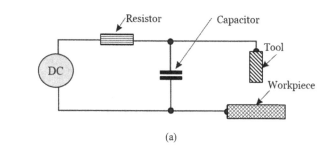

(a)

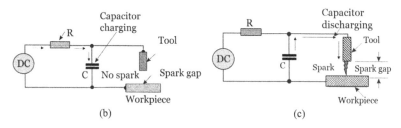

(b)   (c)

FIGURE 2.17
Circuit diagram of (a) general RC circuit, (b) during pulse OFF time, and (c) during pulse ON time.

- The pulse duration and pulse interval time cannot be varied.
- Inability to recover the dielectric strength leads to flow of discharge current through the existing plasma channel, and the capacitor fails to charge during this period. It will cause serious thermal damage on the workpiece surface.

2.4.2.2 Alternating Current Phenomena in RC Circuits

Figure 2.18i shows an ideal pulse shape of the voltage and current signal. However, the real scenario is more complex and interesting. Figure 2.19a shows a more realistic representation of the pulse generator in which the parasitic inductance and voltage measurement points are included (Yang et al., 2018a). In Figure 2.19b, a voltage and a current signal corresponding to the RLC circuit are shown (Yang et al., 2018a). In the diagram, u_{Ce} is the voltage measured across the capacitance C_e which is supplied by the voltage source U. The stray capacitance is indicated as an additional capacitance C_p, and the parasitic inductance and resistance of the circuit wires and voltage measuring probe as L_p and R_p, respectively. In this RLC circuit, the inductance and capacitance produce a resonance during the discharge duration, which pushes the voltage across the capacitor (u_{se}) into a negative value towards the end of a normal discharging cycle. This is followed by a high-frequency resonance in the circuit, and the gap voltage falls to a large

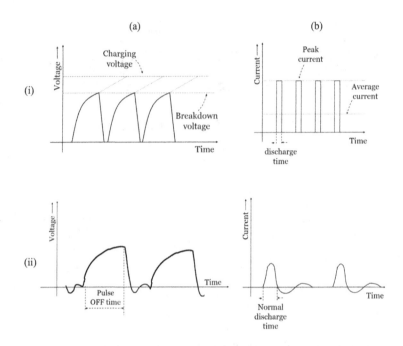

FIGURE 2.18
(i) Ideal and (ii) realistic pulse waveforms of (a) voltage and (b) current signal in RC-based EDM circuits.

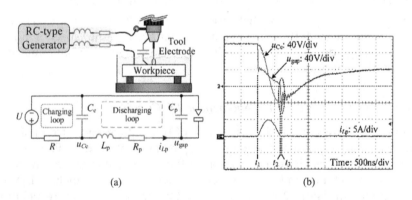

FIGURE 2.19
(a) Realistic representation of the RC pulse generator by including the parasitic inductance in the circuit and (b) voltage and current signals showing the effect of parasitic inductance (Yang et al., 2018a).

negative value. This negative voltage between the gap may reach up to −60 V (Yang et al., 2018a). As the dielectric in the gap may not regain its strength in this short time, an arc discharge will occur. This subsequent discharge after a normal discharge period due to the negative gap voltage is coined as alternating current phenomena.

The reversed discharge phenomena and the heating effect produced by it in the machine zone are analyzed by Albinski et al. (1996). The "ringing" phenomena in micro EDM is characterized by Kao and Shih (2006), and correction models are proposed. The discharge energy of the reversed discharge (secondary discharge wave) is found to be smaller than the normal discharge (primary discharge wave) (Tomura and Kunieda, 2010). The effect of the secondary discharge wave in the material removal mechanism of the tool and workpiece is investigated by Qian et al. (2015). The alternating current effect increases the MRR and tool wear rate. Yang et al. (2018b) studied the alternating current phenomena and reported that the alternating current could enhance the machining efficiency as it consumes energy from the main capacitor. The importance of incorporating the characteristics of alternating current in the process models and servo control models in micro EDM is emphasized in their findings. Alternating current flow is analyzed with pulse counting method by Wang et al. (2016). An electric circuit model including the parasitic inductance and capacitance is developed, and theoretical analysis of the discharge process is conducted by Hua et al. (2018). Figure 2.20 shows the alternating current phenomena and the actual phases of current flow during EDM.

2.4.3 Comparison of RC and Transistor-Based Pulse Generators

Jahan et al. (2009b) conducted a series of experiments to compare the performance of RC pulse generators and transistor-based pulse generators during micro EDM drilling of tungsten carbide. Their major findings are listed below:

1. The quality of microholes fabricated with RC pulse generator is superior in terms of the absence of burr-like recast layer at the hole entrance and better circularity (Figure 2.21).

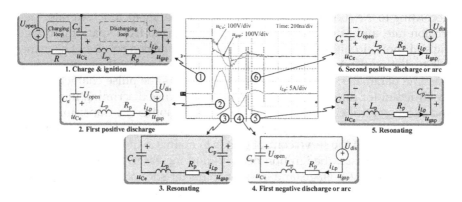

FIGURE 2.20
Alternating current phenomena and the different phases during machining (Yang et al., 2018a).

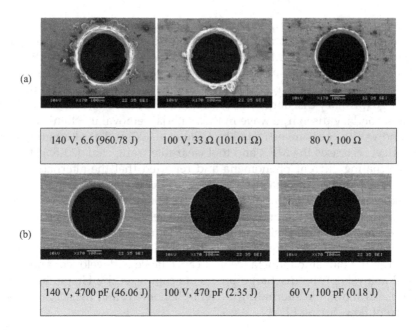

FIGURE 2.21
Entrance section of the microholes machined by micro EDM using (a) transistor-type and (b) RC-type pulse generators (Jahan et al., 2009b).

2. The spark gap is comparatively small in RC-type pulse generators compared to transistor type (Figure 2.22i).

3. The difference between tool entrance diameter and exit diameter is less in micro EDM with relaxation circuit (Figure 2.22ii).

2.4.4 Servo Control Based on Pulse Discrimination

According to the concentration of debris in the spark gap, tool feed rate, tool erosion rate, and MRR, the gap state can be changed from sparking to arcing and short-circuiting. The process monitoring and control circuit have the role to distinguish between different gap states and adjust the servo feed or machine parameters to retain the normal state. As the pulse characteristics are different for different pulse generators, the pulse discriminating circuit and servo control circuit also differ from one another. In transistor-based circuits, ignition delay is used as the gap state monitoring parameter. One of the major aims of the servo control unit is to maintain the spark gap. As the spark gap increases, the ignition delay period increases. By understanding the ignition delay, voltage, and current characteristics of different types of discharges (Figure 2.23), the gap state can be predicted during machining, and the tool feed is adjusted accordingly to avoid undesirable discharges (Kunieda et al., 2005).

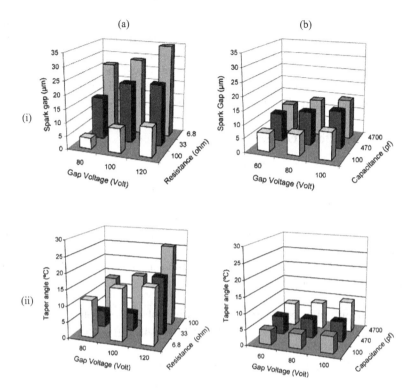

FIGURE 2.22
Comparison of (i) spark gap and (ii) taper angle during EDM microdrilling using (a) transistor-type and (b) RC-type pulse generators (Jahan et al., 2009b).

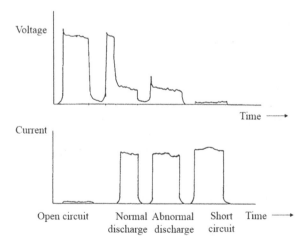

FIGURE 2.23
Pulse characteristics of different types of discharges in transistor-type pulse generator (Snoeys et al, 1983).

In an RC circuit, the average gap voltage/current is used to monitor the gap state. The average voltage input is then compared with the servo reference voltage to adjust the tool feed (Liao et al., 2008). However, for other pulse generator circuits, dedicated servo control systems are developed (Han et al., 2004; Kunieda et al., 2007; Cheong et al., 2018; Yang et al., 2018b). Figure 2.24 shows the method to differentiate the type of discharge by analyzing RC pulse signals.

2.4.5 Modified Pulse Generators

2.4.5.1 Transistor-Based Isopulse Generator

One of the major drawbacks in using transistor-type pulse generators is the formation of nonuniform pulses due to delay in signal transmission. To generate pulses with a similar pulse duration (isopulses), Han et al. (2004) developed an isopulse generator circuit as a modified form of transistor circuit, as shown in Figure 2.25.

Han et al. (2004) identified time constants in the following subsystems as sources of delay in transistor circuit,

- Voltage attenuation system: attenuation of high-voltage pulses to low voltage to prepare the input to digital pulse discrimination and voltage control circuits.
- Pulse control system: controlling the discharge ON–OFF periods.
- Insulation circuit: restricts the current flow into control circuits.
- Gate circuit of FET for discharge current control.

To cut off the power immediately with respect to the gap state conditions, the time delay inherent in the transistor circuit has to be reduced. During a roughing operation, FET_1 controls the discharge current flow by connecting P_1 and P_2. The gap state is monitored and controlled by pulse control unit and gate control unit. To further shorten the pulse duration, the current flow is cut off without using the pulse control circuit. This is achieved with the help of transistors (Tr_1, Tr_2, Tr_3, Tr_4, Tr_5) connected as a subsidiary circuit as shown in Figure 2.25. Moreover, the need for a voltage attenuation circuit is eliminated by changing the controlling parameter from gap voltage to pulse current readings. Using the two FETs effectively concerning machining conditions helps to cut off the current flow instantaneously. This ensured the generation of pulses with the same duration. Figure 2.26 shows a pulse signal with uniform pulses generated by the Tr-isopulse generator (Han et al., 2004). Muthuramalingam and Mohan (2014) evaluated the performance of an isopulse generator by comparing the surface roughness and machining rate of conventional transistor circuits. Surface quality and MRR are improved by the use of a modified pulse generator circuit. Discharge current and duty factor are recognized as the most influential parameters during machining. Isopulse generator-based EDM process is used to make micropits of the hip implant (Mahmud et al., 2012).

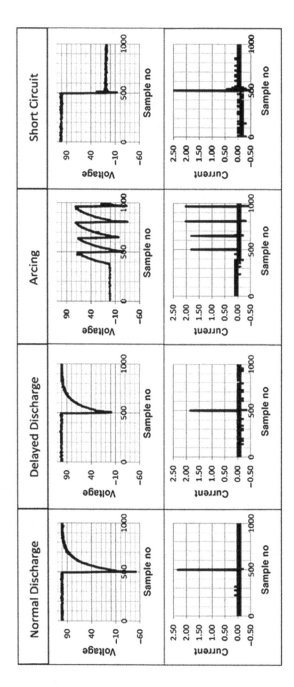

FIGURE 2.24
Pulse characteristics of different types of discharges in RC-type pulse generator (Yeo et al., 2009b).

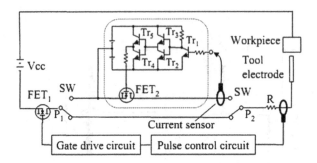

FIGURE 2.25
Circuit diagram of transistor-type isopulse generator for micro EDM (Han et al., 2004).

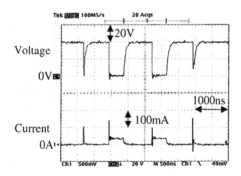

FIGURE 2.26
Uniform pulses generated by employing transistor-type isopulse generator for micro EDM (Han et al., 2004).

2.4.5.2 Smart EDM Generator

An intelligent pulse generator that combines pulse discrimination, pulse counting, and pulse control systems in a single system is developed by Giandomenico et al. (2018). This "all in one" system integrated the voltage measurement, gap state analysis, linear current shape generation, machining control, tool wear assessment, and CNC system in a single circuitry. A constant tool feed rate is used during machining, and the spark gap control is achieved by continuously updating machining parameters and varying the spark frequency. The topology of a smart generator circuit with high-speed metal-oxide semiconductor field-effect transistor (MOSFET) and pulse inversion circuit is shown in Figure 2.27.

2.4.5.3 Bipolar Pulse Generator

A pulse generator that makes use of alternating current phenomena is developed for micro EDM (Yang et al., 2018b). It is capable of realizing a micro EDM process with better MRR and reduced tool wear compared to the conventional unipolar pulse generators. Full bridge circuit with voltage control

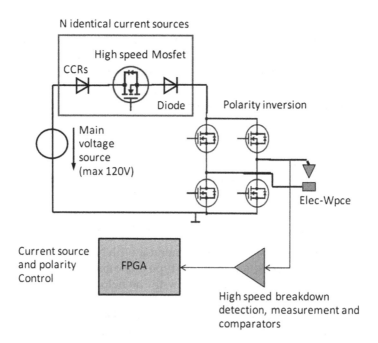

FIGURE 2.27
The topology of the circuit of a smart EDM generator (Giandomenico et al., 2018).

during ignition period and current control during the discharge period generate positive and negative pulses with a fixed discharge frequency and constant energy consumption per cycle. These uniform pulses are capable of enhancing the quality of machining. Figure 2.28 shows the pulse generator topology at two different pulse conditions. During the normal discharge period or positive pulse period, the transistor T_4 is connected to the circuit and T_2 remains open. The pulse generator is then controlled by the buck converter circuit consisting of T_3 and T_4. The discharge energy is controlled by current monitoring to ensure uniform energy per pulse. Similarly, for the

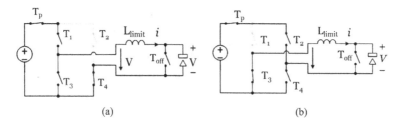

FIGURE 2.28
Principle of the bipolar pulse generator for different phases: (a) positive pulse processing and (b) negative pulse processing (Yang et al., 2018b).

negative pulse period, T_3 is closed and T_1 will be open, which makes a new buck converter with T_2 and T_4 for a negative pulse. The deionization period is controlled by T_{off}. To reduce the pulse duration and achieve high pulse frequency, MOSFET and a diode have been used in the circuit.

2.4.5.4 Hybrid Two-Stage Pulse Generator

To produce ultrashort pulses for micro EDM, Kröning et al. (2015) used a two-stage hybrid pulse generator. The static generator circuit is used for the ignition phase to establish the plasma channel, and the relaxation circuit is used in the power stage for producing the appropriate current waveform. Using this hybrid system, the switching delay can be minimized.

2.4.5.5 Modified RC Circuit with an N-Channel MOSFET Control

Cheong et al. (2018) developed a pulse generator with N-channel MOSFET to prevent the current fluctuations in an RC-type pulse generator. The MOSFET cuts off the current flow before reversal with the help of PWM control. The MOSFET-controlled RC circuit is reported to be capable of reducing the effect of alternating current on tool wear during micro EDM. A general circuit of an RC circuit with MOSFET control is shown in Figure 2.29.

2.4.5.6 Multimode Power Supply

To cope with the requirements of a pulse generator for micro EDM drilling, a multimode power supply system is developed (Li et al., 2013). It consists of

- RC mode
- Controlled resistance mode

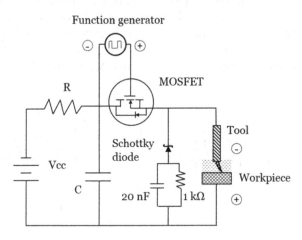

FIGURE 2.29
Circuit diagram of RC pulse generator with MOSFET control (Cheong et al., 2018).

- Grouped controllable resistance mode
- Controlled capacitance mode
- Grouped controllable capacitance mode.

The pulse generation control is done with a microprocessor and complex programmable logic device (CPLD). Single Chip Micyoco (SCM) controls the CPLDs concerning the control signal from the host computer. Pulse power supply system controls the shifting of power supply modes.

2.4.5.7 H-Bridge Generator

To increase the flexibility of pulses regarding duration, amplitude, and sequence, an H-bridge circuit is developed (Giandomenico et al., 2016). A high-speed gap breakdown detection and double polarity ignition are incorporated into the system. High-speed control of pulses is achieved by including an embedded programmable circuit in the pulse generation circuit. There are four modes of operation depending on the generated pulse sequence: (1) positive unipolar sequence, (2) negative unipolar sequence, (3) positive bipolar sequence, and (4) negative bipolar sequence. Using an H-bridge circuit with the transistor control, the pulse sequence and characteristics can be completely configured.

2.4.5.8 Capacity-Coupled Pulse Generator

One of the main hindrances in reducing the effective discharge energy in RC-type pulse generator is the existence of parasitic capacitance or stray capacitance. The capacitance of the voltage-measuring probe and tool feeding arrangement is accountable for increasing the effective capacitance in the circuit. One way to overcome this problem is to develop a noncontact voltage measuring method for gap state detection. Kunieda et al. (2007) developed an RC pulse generator by incorporating a noncontact voltage measuring system based on capacity coupling. Single discharge experiments revealed that capacity-coupled pulse generators are capable of producing craters of 0.43 µm in diameter, which is the smallest crater diameter obtained using an RC pulse generator circuit.

2.4.5.9 Pulse Generator Based on Electrostatic Induction Feeding

To increase the MRR in micro EDM, a method to increase the discharge energy by electrostatic feeding is developed as shown in Figure 2.30 (Abbas and Kunieda, 2016). An additional capacitor is connected to the pulsed power supply which has almost ten times higher capacitance rating than the parasitic capacitance in the spark gap. It is reported that high discharge energy can be supplied to the spark gap if the machining is done at the resonant

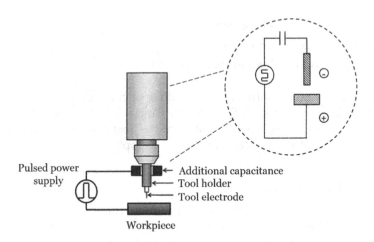

FIGURE 2.30
Configuration and equivalent circuit in the induction feeding method-based pulse generator for micro EDM (Abbas and Kunieda, 2016).

frequency of the circuit, which leads to high MRR. This is achieved with a controlled pulse train method. This generator also makes use of the parasitic inductance effect and bipolar behavior of current flow.

2.4.5.10 PULSAR Pulse Generator

SARIX introduced a new micro EDM controller called PULSAR based on two key technologies, namely, ADP® (Analog Digital Pulse) and DPM® (Direct Pulse Motion). It is claimed to be capable of generating micro- and nanopulses and quick response characteristics towards process control (www.sarix.com).

2.5 Servo Control System

The motion control system ensures that the motion stages follow an intended path and completes the position feedback or velocity feedback loop. Steps in motion control for machine tools can be generally summarized as follows:

- A sensor system generates control signals during the stage motion.
- The control signal is compared with reference signals using a micro-controller circuit.
- A driver system converts the error corrections to appropriate input signals to the motor and synchronizes axes motions.
- The motor controls the movement of stages accordingly.

The servo control system is an essential part of the positioning system of a CNC machine tool. The servo control is usually connected to the servomotor/ linear actuator. A PWM unit manipulates the pulse inputs to the motor concerning the error signal to control the rotation of the motor. In conventional machine tools, one of the error signals to the controller is generated by comparing the exact position of the tool/workpiece concerning the intended location using linear encoders. The error in position is then compensated as the lead screw converts the rotation motion to linear motion and corrects the position of the tool/workpiece. Conventional machine tools' servo control systems mainly focus on the position and vibration control. In micro EDM, apart from position and vibration control, the spark gap between the tool and workpiece has to be maintained within the limit (Jiang et al., 2012b). This extra duty to the servo control system makes the control algorithms more complex. However, being a noncontact process, the effect of cutting force and chatter is negligible, which is advantageous in vibration control. An overview of the servo control system in micro EDM drilling is given in Figure 2.31. For micro EDM drilling, controlling the Z stage will be enough to maintain stable machining. However, in micro EDM milling, contouring, and planetary EDM, all the motions have to be controlled by a specialized servo controller to avoid workpiece/tool damage and occurrence of harmful discharges.

The effective spark gap in micro EDM is affected by the following factors:

1. Material removal from the workpiece
2. Tool electrode wear
3. Presence of debris particles.

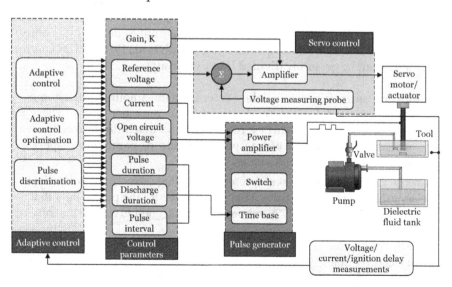

FIGURE 2.31
Overview of servo control/gap state control strategy for micro EDM (Snoeys et al., 1983).

As the material is removed from the tool and workpiece, the spark gap increases. As the maximum voltage gradient (ratio of open circuit voltage to spark gap width) crosses the critical lower limit, the machining process is stopped. To resume the material removal action, the tool has to be fed towards the workpiece to keep the voltage gradient within the allowable range (Kruth et al., 1979b). Measurement of a certain electrical parameter (e.g. average gap voltage) generates a control signal which is further amplified, filtered, converted to digital signal, and sent to the comparator where the value is compared with the reference value of the parameter and the error signal is generated. This error signal helps to manipulate the motor rotation with the help of a motor drive and compensates for the error. The spark gap also depends on the MRR and tool erosion rate. So, the spark gap control without considering what is happening in the machining zone may not be always reliable. To understand the situation in the spark gap, the parameters that affect the gap state have to be constantly monitored. This includes discharge time, discharge current, pulse duration, pulse interval time, open circuit voltage, the polarity of the electrode, dielectric flow rate, etc. Fixing the reference value for the control parameter for maintaining "useful discharge state" is considered as the most important step in motion control. For this, either the mathematical relation between spark gap and the reference parameters must be well known, or the reference value has to be decided by a tuned adaptive control circuit. However, being a stochastic process, the characteristics of EDM are very difficult to predict, and the empirical relationships may lack in accuracy.

Moreover, the process of finding empirical relations demands long experimentation and man hours. In contrast, an adaptive control system (ACS) predicts the trend of machining parameters and makes decision to control the undesired situations during machining (Kruth et al., 1979b). EDM control parameters can be generally divided into two groups:

Group 1: Power parameters—they determine the amount of discharge energy exerted locally on the machining area per pulse. These parameters affect the surface roughness and machining time. They are generally selected offline with the help of parametric experiments to select optimum values.

Group 2: This includes servo reference voltage, electrode retraction length, and tool jump time. This group refers to the parameters that need to be continuously monitored and controlled to ensure machining stability.

Primary control objectives used by different researchers include feed speed (Yeo et al., 2009b), tool downtime (Zhou and Han, 2009), and discharge pulse parameters (Kruth et al., 1979b).

Other control objectives: Apart from spark gap control, adaptive control has to be applied for some other subsystems to ensure that the process is less affected by disturbances (Kruth et al., 1979b).

1. The pulse interval time: As the concentration of the debris particles increases in the dielectric fluid or the depth of machining increases, deionization time required for the dielectric fluid may change. To avoid arcing, the pulse interval has to be controlled and optimized during machining.
2. Dielectric flow rate: To control the debris concentration and the pressure exerted by the fluid on microelectrodes, the dielectric flow rate has to be controlled and optimized.

2.5.1 The Need for Tool Retraction

The speed of machining in micro EDM depends on the chance of occurrence of short circuits during machining. Even though a predefined tool feed has to be given before starting machining, the tool may not be able to advance unless the material is removed from the workpiece surface. Stable machining in EDM refers to the instance where the frequency of normal discharges is more than the frequency of arcs and short circuits. During EDM, as the debris concentration in the machining zone increases, the flushing gets difficult, and the frequency of arc discharges and short circuits increases. The tool has to be periodically retracted from the machining point to facilitate dielectric flushing (He et al., 2010). Maintaining the spark gap in the optimum values and rapid retracting of the tool electrode with the detection of harmful discharges will establish stable machining in micro EDM. Electrode retraction length and jump time (time to retract and advance back to the earlier point) affect the machining time (Zhou and Han, 2009). As the retraction length increases, more time will be needed to bring back the tool to the machining zone. On the other hand, as the retraction length decreases, the arc discharges may not disappear, and the workpiece surface will experience thermal damage. The jump time has to be carefully determined to allow proper dielectric flushing and deionization in the machining zone (Zhou et al., 2018).

2.5.2 Servo Control for the RC Circuit

In an RC circuit equipped with EDM, wires are usually connected between the tool and the workpiece to acquire the voltage/current signals, which are then amplified and filtered. However, in micro EDM, these wires will increase the stray capacitance in the circuit, which effectively increases the discharge energy. To reduce stray capacitance, two resistors are connected to the wires, which reduces the chance of charge flow due to stray

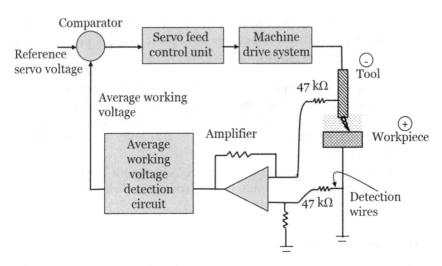

FIGURE 2.32
The servo control system of micro EDM machine tool with RC pulse generators (Vaezi et al., 2016).

capacitance (Vaezi et al., 2016). The servo control circuit consists of voltage/current sensing circuit, amplifier, filter, comparator, feed control circuit, and the servomotor-lead screw mechanism (Figure 2.32). Average gap voltage is generally used as the control spark gap in RC circuits.

2.5.3 Servo Control for the Transistor Circuit

As the stray capacitance between FET and resistor in transistor circuit is itself capable of charging the spark gap, the average gap voltage cannot be used to control the spark gap. So, the average ignition delay time is used for servo control in a transistor circuit (Vaezi et al., 2016). Servo reference voltage is used as the input to the comparator to generate the error signal. Figure 2.33 shows the servo control system designed for a transistor-based pulse generator.

2.5.4 Model of the Gap Control System

A model of the spark gap and reference parameter is developed for servo control (Chang, 2005; Zhou and Han, 2009). In the servo control system, the feedback signals generate the input signals to the CNC interpolator. The interpolator is responsible for separating the feed signals to the different axes for synchronous motion. As the positive feed rate F_r signal is given to the interpolator, it takes the electrode forward. y_t can be considered as the feedback signal and y_r as the reference input. The difference is the error value $e(t)$. The gap controller $H(t)$ determines the feed rate concerning the error $e(t)$ as given in Eq. (2.1). Modeling of the control parameter $u(t)$ is given in Eq. (2.2).

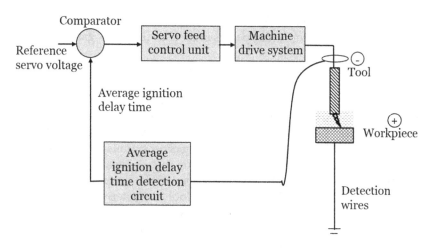

FIGURE 2.33
The servo control system of micro EDM machine tool with transistor-type pulse generators (Vaezi et al., 2016).

$$F_r = H(t) \times e(t) \tag{2.1}$$

The primary control parameter $u(t)$ can be modeled as

$$u(t) = \frac{C}{qB} y_r - \frac{C - A}{B} y_t \tag{2.2}$$

where y_r is the reference value and y_t is the feedback value from the measured parameter indicating the current gap state. A, B, and C are the polynomials used to model the process, and q is the forward shift operator.

As the electrode is fed towards the workpiece surface, the distance from the workpiece surface, $d(t)$, can be calculated by integrating the actual speed. However, the actual spark gap distance $d_e(t)$ is affected by material erosion from the workpiece and tool $d_m(t)$ and disturbed distance $d_d(t)$ due to the presence of debris particles. The equivalent spark gap will be the sum of the actual spark gap, a gap due to material removal minus gap disturbance due to the presence of debris particles, which is explained in Eq. (2.3).

$$d_e(t) = d(t) + d_m(t) - d_d(t) \tag{2.3}$$

As $d_m(t)$ depends on machining parameters and $d_d(t)$ is an unpredictable value showing the stochastic nature of the EDM process, maintaining the actual spark gap within the desired range is complex and challenging. The parameters that are used to define the control function have to be continuously updated to incorporate the disturbances in the system. This procedure is known as adaptive tuning of the controller. As direct measurement of $d_e(t)$ is difficult, it is represented as a function of some electrical parameter of EDM. Change in that parameter is mapped as change in the spark gap distance.

2.5.5 Adaptive Control Systems in EDM

ACSs adapt to the timely variations in the process characteristics and provide real-time control to the process (Hashim et al., 2015). ACS was first introduced in the 1970s and was popularly employed to regulate the axes movements of EDM machine tool concerning error signals. The error signals are often compared with the reference value to compensate for the differences. ACS is capable of modifying the response characteristics concerning the unexpected variations in the machine tool characteristics. For this, the process model and controller parameters have to be accurately defined and modified accordingly. Therefore, the control system of ACS is comprised of a self-tuning regulator that modifies the control parameters concerning disturbances. A method of adaptive control employing a self-tuning regulator developed by Zhou and Han (2009) is shown in Figure 2.34. In this, a minimum variance strategy is implemented for self-tuning the control parameters. The input signals are measured in specified intervals, and the control parameters are updated accordingly. The gap state is controlled with the help of changing the tool downtime, which is a function of the control parameter $u(t)$. The control parameters determine the values of polynomial coefficients, and the model is updated according to machine disturbances.

Kruth et al. (1979a) developed an ACS to regulate machining stability. The pulse characteristics can be used to evaluate the gap state during EDM. Dauw et al. (1983) developed a pulse discriminator that is capable of identifying normal discharges, arcing discharges, and short circuits by analyzing the pulse characteristics. By effectively modeling the pulse discriminators, the tool wear and material removal can be predicted by analyzing the time history measurements. Snoeys and Cornelliss (1975) used ignition delay time

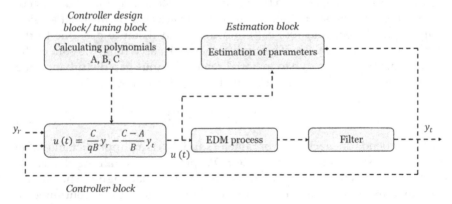

FIGURE 2.34
The principle of an ACS for EDM servo control (Zhou and Han, 2009).

for EDM pulse discrimination. According to Yeo et al. (2009b), peak current is believed to be giving more realistic data regarding the machining characteristics, including the amount of discharge energy exerted by a single pulse. The discharges are discriminated as normal discharge, delayed discharge, arc discharge, and short circuit with the help of peak current. Servo control is established by calculating the frequency of different discharges over time. An adaptive control based on a specific model that represents the gap states as a function of time ratios of different discharge states is then developed as an alternative control strategy (Rajurkar et al., 1989). During this time, servo voltage is used as the gap control reference input. Researchers tried to develop gap state monitoring with the help of RF signals (Rajurkar and Royo, 1989). A fuzzy control system has been utilized by Boccadoro and Dauw (1995) to control EDM process parameters. Adaptive control based on tool jump time is realized by Wang et al. (1995).

Most of the ACSs is designed for transistor-type pulse generators so that applicability of those systems for micro EDM is questionable. In micro EDM, RC circuits are generally used to generate discharge pulses, where the ignition delay phenomenon is not visible to characterize the gap state. Moreover, in RC circuits, the discharge time is very short so that the data acquisition system has to be highly responsive to acquire the smallest deviation and disturbances in the system. Yeo et al. (2009a) successfully developed a dedicated ACS for micro EDM with pulse discrimination system based on the RC circuit. Self-organizing fuzzy sliding mode control strategy has been utilized by Yan (2010) for adaptive control. Among these controller systems, the fuzzy control system is most adaptable to the unpredictable machining environments (Boccadoro and Dauw, 1995). With the help of a neural network or genetic algorithm-based optimization methods, the fuzzy logic adaptive controller can be effectively optimized (Behrens and Ginzel, 2003). Figure 2.35 shows the principle of a fuzzy logic controller for micro EDM. Table 2.3 summarizes various literatures on adaptive control for EDM.

2.5.6 PID Control

The proportional-integral-derivative (PID) controllers are generally used to control the servo movements by employing a mathematical model to predict and compensate the error in positioning. In EDM, PID controller has been widely used for position control and gap state control (www.compumotor.com). The general control strategies based on mathematical models are as follows:

Proportional (P) controller (K_p): This is often called the "present" controller. Here, the error signal is proportional to the feedback value. The presence of mass elements in the positioning systems such as the stages has inertia, and the proportional control will result in oscillations and unstable machining. It reduces rise time, but the steady-state error persists.

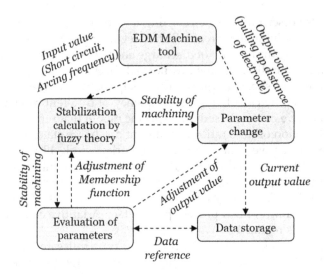

FIGURE 2.35
Self-adjusting fuzzy control strategy (Kaneko and Onodera, 2004).

TABLE 2.3

Summary of Literatures on Adaptive Servo Control

Authors	Adaptive Servo Control Approach	Remarks
Kruth et al. (1979a)	Steepest ascent path algorithm	Pulse efficiency (normal discharge time/total pulse duration) is used as the control parameter. Pulse OFF time and the servo reference voltage are changed for maximum pulse efficiency during machining
Wang (1988)	Self-tuning-based dynamic modeling approach	Adjusting servo reference voltage and discharge interval based on pulse discrimination
Rajurkar et al. (1989)	Using a simplified reference model by establishing a linear relationship between the output parameters to be stabilized and the gap parameters	The difference between process variables and the model outputs is minimized for machining stability
Rajurkar et al. (1990)	A real-time stochastic model for direct feed control	Discharge time ratios have been monitored to directly control the tool feed without relying on the servo reference voltage
Wang et al. (1995)	Adaptive tool jumping control system based on discharge time ratio	Increases the efficiency of deep hole EDM drilling by optimizing jump downtime of the tool

(*Continued*)

TABLE 2.3 (*Continued*)

Summary of Literatures on Adaptive Servo Control

Authors	Adaptive Servo Control Approach	Remarks
Boccadoro and Dauw (1995)	Fuzzy logic control to optimize the reference values	The machining stability is controlled by predicting and avoiding harmful discharges with the help of fuzzy logic system
Behrens and Ginzel (2003) and Fenggou and Dayong (2004)	Self-learning algorithm based on artificial neural network	Machining rate is improved by controlling the gap state and pulse efficiency with the help of continuously learning-optimizing neural network
Suganthi et al. (2013)	Adaptive neurofuzzy model	Modeling to predict the output parameters for stabilizing the MRR and tool wear.
Xi et al. (2017)	Gap voltage estimation with the help of Kalman filter	Treating gap voltage as the sum of colored noise through linear filter and measurement noise. A state space model is used to filter out the noises in gap voltage measurement for servo feed control
Mahyar et al. (2018)	Discharge pulse durations	Self-tuning adaptive controller based on MIT rule. Doubling the machining rates by stabilizing the discharges

Integral (I) controller (K_i): The integral control can increase the stability of moving systems by reducing the rising time and removing the steady-state error. This is capable of reducing the steady-state error, but it is not suitable for transient responses.

Derivative controller (K_d): This is capable of predicting the future from the trend of error signals. This increases the stability of the system and reduces overshoot, but it is not capable of controlling the rise time.

Proportional-integral (PI) controller: It has been used in multiaxis with individual system dynamics for synchronous control. Velocity control is also employed in servo system where nonlinear friction component is considered as a disturbance. Nonetheless, an integral controller causes a limit cycle around the intended position during point-to-point control, and the tracking error gets magnified during the reversal of motion direction.

Proportional-derivative (PD) control: Lee and Tomizuka (1996) used the PD controller to increase the stability of positioning systems in EDM. Kumar et al. (2008) conducted a comparative study of these controllers to judge the effectiveness of the system towards process stability.

PID controller: This is the most commonly used controller method in EDM. A variation of the system known as differential PID controller (DPID) also proved its capability to improve the machining performance in die-sinking EDM

(Chung et al., 2009). However, delay, nonlinearity, and lack of dynamics of the tuning process make the control process complex and difficult to optimize the control parameters (Jaen-Cuellar et al., 2013). This demands some dedicated optimization techniques to be coupled with the controller system to optimize the PID parameter values (Hashim et al., 2015). As the process is stochastic in nature, the optimization process must be dynamic to adapt to the variations or disturbances in the system. This includes reducing steady-state error, reduces settling time, and is capable of handling transient responses. The performance of the motion control system depends on the controller parameters obtained by tuning. The PID control parameters can be tuned using empirical models, analytical models, and soft computing models (Jaen-Cuellar et al., 2013).

General representation of the model for PID control is given in Eq. (2.4).

$$u(t) = K_p e(t) + K_d \frac{e(t)}{dt} + K_i \int e(t)dt \qquad (2.4)$$

where $u(t)$ is the parameter used to control the drive and $e(t)$ is the error between the feedback reading and reference value. For conventional machine tools, K_p, K_d, and K_i are fixed according to the experience of the operator or with preliminary experimental analysis. However, a stochastic process like EDM cannot be controlled in that way. The servo reference voltage and the control parameters have to be updated concerning changes in machining conditions for precise spark gap control. Moreover, in micro EDM, as the servo controller has to maintain very small spark gap for gap state control and the spark duration is as small as nanoseconds, servo control based on fixed models becomes less efficient (Yang et al., 2011). Table 2.4 shows the general methods of adaptive control optimization.

2.5.7 Pulse Discrimination for Gap State Control

The discharge states are generally classified as useful and harmful discharges. The useful discharges are those uniform random discharges responsible for smooth material removal during EDM. Harmful discharges damage the workpiece surface if not controlled effectively. The discharge states can also be classified as open circuit, normal, arc, and short circuit. In arc discharges, the discharge energy channelizes to a specific point instead of uniformly distributing over the workpiece surface. This happens due to the increased concentration of debris particles in the machining region. In short circuits, the tool and workpiece come in contact.

One of the important steps in controlling the stability of EDM process is to understand the gap state. The pulse train contains the data that can explain what is happening in the spark gap. Among the various signals that can be acquired from the spark gap during machining, average gap voltage is considered as one of the simplest and effective parameters to understand the gap state. This is the reason behind the popularity of using average gap voltage for servo feed control in an EDM process. However, completely relying

TABLE 2.4

General Methods of Adaptive Control and Adaptive Control Optimization

Authors	Controller/Optimization Type	Remarks
A General Comparison of the Servo Control Systems and Their Tuning Methods		
Jain and Nigam (2009)	PD-PI + GA/PD-PI + PSO	Gain parameters K_p, K_i, and K_d get higher values during PSO optimization compared to GA, but PSO is faster than GA. Easy implementation and more efficient global searching abilities makes PSO a more reliable optimization method compared to GA
Thomas and Poongodi (2009)	PID + GA	Less response time, less error, less rise time, and settling time for the controller
Allaoua et al. (2009)	Fuzzy + PSO/PID + PSO	PID controller with PSO shows faster servo control compared to the fuzzy-PSO system
Comparison of the Controller System for EDM		
Mahdavinejad (2009)	Fuzzy + GA + neural network	EDM experiments show the controller can improve the MRR by up to 32.8%
Chang (2007)	PI + GA	A material removal process in EDM with discharge pulses can be modeled with first-order dynamic process with nonlinear and time-varying perturbations
Andromeda et al. (2012b)	PID + differential evolution	PID controller with DE optimization shows better performance in attaining the optimized parameter with minimum fitness function and less processing time
Chang (2005)	PD + GA + mixed H_2/H_∞ optimization	This shows an optimal tracking performance and tolerates the nonlinear time-dependent feedback signals
Kao et al. (2008), Yan and Chien (2007), Byiringiro et al. (2009), Liu and Zeng (2011), and Zhang (2005)	Fuzzy logic control	Fuzzy logic systems are capable of dealing with the nonlinear stochastic behavior of the discharge pulses in micro EDM. Fuzzy logic is generally used to discriminate the gap states and predict the harmful discharges and control parameters to avoid them

(Continued)

TABLE 2.4 (*Continued*)

General Methods of Adaptive Control and Adaptive Control Optimization

Authors	Controller/Optimization Type	Remarks
Wang et al. (2004) and Yan and Fang (2008)	Fuzzy-neuro controller + GA	Fitness function based on Lyapunov design approach. A supervisory controller incorporated to ensure the stability of the machining system
Zhang et al. (2012)	Two-stage controller based on type 2 fuzzy logic	Two-stage type extension is performed for input and output membership functions. This controller can eliminate the shortcomings of type-1 fuzzy control, particularly at the deteriorating environment
Andromeda et al. (2012a)	PID + particle swarm	Optimum gap achieved with different combinations of the particle swarm optimized gain parameters. PD controller showed a minimum settling time

GA, genetic algorithm; PSO, particle swarm optimization; PD, proportional derivative; PI, proportional integral.

on the data based on average gap voltage for servo control can result in unexpected errors (Liao et al., 2008). Servo control based on reference servo voltage cannot predict or act before the occurrence of harmful discharges. However, the pulse discriminators can identify the trend and act little earlier. But, understanding the behavior of pulses in micro EDM is very challenging as the pulse duration and pulse energy are very small, and small errors will have a large impact on machining stability. The first attempt to develop a discriminating pulse system for micro EDM is made by Liao et al. (2008). A differential probe is used to measure the gap voltage. The voltage train characteristics (voltage and pulse duration) are used to distinguish effective arc, normal discharge, transient short circuit, and complex discharges. Voltage and pulse duration levels are classified as low, medium, and high. Generally, during the charging time of capacitance, the voltage gradually increases. As the discharging takes place, the voltage drops to zero or below zero because of electrotransitivity. The discharge current flows through the gap and reaches a peak. So, the discharge interval/discharge duration will be equal to the charging time/discharging time of the capacitor. These discharges are called normal discharges. The dielectric fluid with a high breakdown voltage or high resistance may result in charging of the capacitor before reaching zero voltage. Liao et al. (2008) called these discharges as effective arc discharges where the voltage never drops to zero and the pulse discharge time will be shorter. Transient short circuits happen due to increasing concentration of debris in the machining area. In these discharges, the voltage drops to zero, but the discharging time seems to be longer than the normal discharge. This alters the stability of the machining system. A "complex pulse" with steep valleys and flat bottom in the voltage pulse train happens due to high debris concentration, and the process becomes totally unstable and unpredictable. Liao et al. (2008) utilized this correlation between gap voltage and pulse duration to understand the type of discharges during micro EDM. Table 2.5 summarizes different pulse discrimination strategies used for EDM and micro EDM. Apart from servo control, pulse discriminators are widely used in micro EDM for online tool wear compensation.

2.6 Dielectric Supply

Dielectric fluid behaves as an insulator at normal conditions and as a conductor when a potential greater than the breakdown voltage is applied. The main functions of dielectric fluidic are as follows:

- Act as a medium for controlled electrical discharges in EDM
- Act as a quenching medium to cool and solidify gaseous EDM debris

TABLE 2.5

Pulse Discrimination Strategies for Gap State Control in EDM

Authors	Discrimination/ Control Parameter	Remarks
Process Control Based on Pulse Discrimination in EDM		
Snoeys and Cornelliss (1975)	Ignition delay	Summing up the discharge duration per second and correlating with MRR to find effective discharge duration
Dauw et al. (1983)	Voltage rise time Ignition delay time Discharge duration	Correlating the effective pulse discharges with tool wear to identify the unstable machining conditions and machining inaccuracies
Weck and Dehmer (1992)	Ignition delay time Fall time	Pulse discrimination based on chaos theory
Snoeys and Dauw (1980)	Voltage threshold Pulse OFF time	Steepest ascent approach and a sectioning strategy were used to control the process
Blatnik et al. (2007)	Discharge current	Surface current density is used to identify different regimes based on the percentage of harmful arc discharges
Zhou et al. (2008), Ter Pey Tee et al. (2013), and Blatnik et al. (2007)	Gap voltage Gap current	Fuzzy logic is used to discriminate the pulses
Cogun (1988)	Voltage and time characteristics of voltage pulse trains	Time lag duration of discharges are used to evaluate the MRR and tool erosion rate
Coğun and Savsar (1990)	Ignition delay time	Ignition delay probability distribution is used to distinguish various gap states. A regression model connecting MRR and segmented ignition delay time probability distribution is formed for gap state analysis
Rajurkar et al. (1989)	Open circuit voltage Discharge voltage Discharge current	Comparing the spark gap reading with predetermined values of the gap parameters, the discharges are classified as a normal spark, transient, unstable arc, harmful arc, and short circuit
Yu et al. (2001)	Discharge voltage Discharge current Voltage coefficient Current coefficient	A wavelet transform is applied to the feedback signals to retrieve the gap state
Tarng et al. (1997) and Tarng and Jang (1996)	Discharge voltage Discharge current	Intelligent gap state discrimination with fuzzy interface machines and genetic algorithm
Ter Pey Tee et al. (2013)	Gap impedance	Indirect measurement of gap impedance from voltage and current signals and setting threshold voltage and threshold current to differentiate the discharge states

(Continued)

TABLE 2.5 (*Continued*)

Pulse Discrimination Strategies for Gap State Control in EDM

Authors	Discrimination/ Control Parameter	Remarks
	Process Control Based on Pulse Discrimination in Micro EDM	
Liao et al. (2008)	Voltage threshold Pulse duration	Pulse discrimination dedicated for micro EDM based on RC circuits
Nirala and Saha (2016) and Nirala et al. (2017)	Percentage of open circuit voltage	Virtual signal-based simulation for pulse discrimination
Jiang et al. (2012a)	Voltage signal	A wavelet transform method is used to self-tune the servo reference voltage to reduce the percentage of harmful discharges in micro EDM drilling
Kao and Shih (2006) and Kuo et al. (2002)	Voltage rise time Discharge duration	An algorithm based on Dauw et al. (1983) to identify the pulse states in micro EDM. A new model based on serial RLC circuit is realized. Predischarging current phenomena have been reported
Jia et al. (2010)	Gap voltage Gap current	Empirical mode decomposition method is used for pulse discrimination. Gap states in micro EDM are predicted by multisensor data fusion and fuzzy logic
Nirala and Saha (2015)	Percentage of open circuit voltage Gap voltage Gap current	A fusion of voltage and current characteristics are used to predict the gap states in micro EDM
Yeo et al. (2009a)	Discharge duration Pulse on dividing value	Acquisition length is fixed as the sum of discharging time and pulse OFF time. Counting different discharge states is carried out to control the servo feed
Liu et al. (2014)	Voltage and current Feed velocity	Prediction of gap states based on calamities gray prediction theory. Output velocity is used as the predictive target. Electrode pullback velocity is considered as the calamity value. A fusion strategy has been applied to adjust the output of the fuzzy controller (velocity) using the inputs from pulse discriminator
Yang et al. (2010)	Gap voltage Gap current	Prediction and control of the discharge states using simulations based on MATLAB fuzzy module
Byiringiro et al. (2009)	Gap voltage Gap current	Fuzzy logic-based controller to control the gap by taking the classified pulses as fuzzy sets
Mahardika and Mitsui (2008)	Discharge energy per pulse	Using histograms, the discharge pulse number and energy for different discharge states are found out
Modica et al. (2014)	Gap voltage Gap current	Correlating pulse distribution and machining performance to predict stable machining conditions in real time

- Wash away the solidified debris from the discharge gap to filter it out from the system
- Act as a heat transfer medium to take away the heat generated in tool and workpiece due to electrical discharges.

The main properties to be considered are as follows:

1. Viscosity: Low viscous fluids exhibit good flushing abilities and flow easily in the small spark gap in micro EDM.
2. Flashpoint: The flashpoint for EDM dielectric fluids is generally in the range of 70°C–125°C, and it has to be kept higher.
3. Dielectric strength: In EDM, dielectric strength is often denoted by the ionization potential or breakdown voltage of the liquid. Keeping the breakdown voltage high will increase the machining time, and keeping it low will induce frequent arcing during machining. However, the dielectric strength is also dependent on various external parameters including the concentration of debris in the fluid.
4. Pour point: This is a temperature below which the fluid will not flow freely. Even though this property will not affect the machining quality directly, the pour point of the dielectric has to be kept high to avoid storage-related problems.
5. Volatility: The volatility of the dielectric fluid has to be low.
6. Oxidation stability: Dielectric fluid must possess high stability towards oxidation.
7. Effects on the skin: The dielectric fluid must not make skin reactions, and it should have fewer solvent properties.

   However, dielectric fluid with low viscosity is more likely to have a low flash point and high volatility. Dielectric fluids with high flash point may have high viscosity. Therefore, the selection of dielectric fluid has to be done with care. EDM3 oil is one of the popular choices as a dielectric fluid for micro EDM variants.

Life of the dielectric fluid can be accessed by

- Change in oil color: The color of dielectric fluid changes with excessive moisture absorption, high debris deposition, and excessive carbon deposition.
- Increase in viscosity
- Change in odor
- Increase in cutting times
- Increase in the number of unstable arcs and short circuits
- Constant deterioration of machined surface quality.

2.6.1 Common Dielectric Fluids and Properties

Among the dielectric fluids used in EDM, hydrocarbon oil and deionized water are the most popular choices in micro EDM (Jahan et al., 2014).

2.6.1.1 Hydrocarbon Oil

The dielectric characteristics of hydrocarbon oils are much stable during EDM with rapid changes in temperature and increase in debris concentration. This makes hydrocarbon oils a better choice for micro EDM (Jameson, 2001). The main characteristics of hydrocarbon oils while using for micro EDM include

- The absence of electrolytic damage
- High-quality machined surfaces
- Low MRR
- The machined surface becomes more brittle and hard
- Carbon depositions during machining reduce the number of stable discharges and MRR
- CO and CH_4 are liberated during machining with kerosene as the dielectric, which makes the machining environment toxic
- Used for die-sinking micro EDM, micro EDM milling, and micro EDM drilling.

2.6.1.2 Deionized Water

The main characteristics of deionized water are as follows:

- High MRR
- Susceptible to electrolytic damage due to electrolysis
- May induce corrosion in machine parts and workpiece. To avoid electrolysis and corrosion, special power supplies have to be used (Chung et al., 2011)
- Comparatively low-quality machined surface due to electrolytic corrosion.
- High tool wear rate
- Mainly used for wire EDM.

Kibria et al. (2010) and Tiwary et al. (2018) conducted studies to reveal the effect of dielectric fluids (deionized water, kerosene, DEF-92 EDM oil, powder-mixed EDM oils) on the performance of micro EDM drilling on titanium alloy Ti-6Al-4V. They concluded that deionized water provides high MRR during EDM due to the formation of titanium oxide layers with a low melting point. For kerosene, layers of carbide forms of tungsten make the material removal difficult. The decomposition of kerosene to carbon particles and their deposition on the tool electrode surface reduce tool wear. Drilled microholes show comparatively more deviation in diameter when machining with deionized water at high discharge energy. However, at low discharge energy, kerosene shows high overcut compared to water. White layer (recast layer) formation was less for deionized water compared to kerosene.

2.6.2 Special Dielectric Fluids

Yan et al. (2018) used oxygen-assisted nitrogen plasma jet (NPJ), nitrogen–oxygen-mixed plasma jet (NPJ + O_2), and compressed air-assisted (CA) NPJ in order to increase the stability of micro EDM process. Manivannan and Pradeep Kumar (2018) conducted experiments with cryogenically cooled micro EDM set up as shown in Figure 2.36 and reported that the MRR is improved up to 62% and the surface roughness to 36%.

2.6.2.1 Powder-Mixed Dielectric Fluids

Many researchers experimented on powder-mixed dielectric fluids to increase the surface quality of EDM-machined microstructures. Table 2.6 consolidates some important outcomes in this area. More information on this topic is provided in chapter 6.

FIGURE 2.36
Experimental setup of cryogenically cooled micro EDM (Manivannan and Pradeep Kumar, 2018).

TABLE 2.6

Major Research Works in Micro EDM with Powder-Mixed Dielectric Fluids

Authors	Powder Material	Findings
Yeo et al. (2007b)	SiC	Craters with shallow and constant depth are produced during machining with powder-added dielectric fluid.
Chow et al. (2008)	SiC	Distribution of discharges resulted in the more uniform surface. The presence of powder particles resulted in larger slit width and tool wear.
Kibria et al. (2010)	B_4C	Adding B_4C to kerosene does not make much difference in MRR. B_4C-mixed deionized water performed well with high MRR due to efficient distribution of discharge. Recast layer thickness also reduced.
Prihandana et al. (2011)	Nanographite powder	Machining time is reduced by 35% in the presence of graphite nanoparticles because of the stable machining process and increased spark gap. Microcracks on the surface are eliminated with the addition of nanoparticles due to reduced discharge power density.
Prihandana et al. (2014), Prihandana et al. (2009a,b)	MoS_2 powder	Maximum MRR achieved at a powder concentration level of 10 g/l.50 nm size particles gives the highest MRR compared to particles with 10 nm and 2 μm.
Tan and Yeo (2011)	SiC	15%–35% reduction in recast layer is observed while machining with Idemitsu Daphne™ Cut HL-25 dielectric fluid with 0.1 g/l of powder concentration.
Cyril et al. (2017)	Al, Gr, SiC	Tool wear decreased during machining due to enhanced heat dissipation in the presence of powder particles. Material migration observed on the workpiece surface due to implosion gases in the interelectrode gap.
Elsiti and Noordin (2017)	Fe_2O_3	MRR increased with increase in powder concentration. However, extreme powder concentration resulted in powder settling problem and bridging effects.
Modica et al. (2018)	Garnet	Deionized and tap water fluids mixed with Garnet showed deposited layers.
Tiwary et al. (2018)	Cu powder	At higher peak current, powder-mixed deionized water showed minimum overcut and improved circularity. The white layer is not observed during machining with Cu powder-mixed dielectric fluid.

2.6.3 Dielectric Supply System

The dielectric recirculation circuit is comprised of the work tank, overflow regulator, primary filter, dielectric carrying pipes, secondary filter, dielectric tank, dielectric pump, flow regulator, and nozzle. The main function of the circuit is to ensure a continuous flow of dielectric fluid with minimum fluid loss. The debris particle is filtered out using primary and secondary

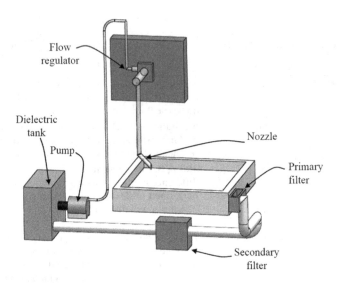

FIGURE 2.37
The dielectric fluid delivery system.

filters. The main aim of employing a primary filter is to avoid the entry of large foreign particles, such as chips, entering the circulation system. Flow regulator controls the discharge volume, and the nozzle controls the jet diameter and pressure. Intelligent flow regulators can be employed for online control of the dielectric fluid pressure (Kruth et al., 1979b). Similar to conventional EDM process, the dielectric flow can be internal or external. However, employing internal flushing in a microelectrode is challenging. Figure 2.37 shows a general outline of the dielectric circulation system used in micro EDM.

2.7 Measurement Systems

The primary role of the measurement system is to monitor the tool electrode condition and dimensions. On-machine fabrication of the tools demands proper metrology systems to check the tool status regularly. Generally, a charge coupled device (CCD) camera with a light source connected to an image-processing system is used to access the dimension of microfeatures on tool/workpiece electrodes (Guo et al., 2006). A shadow image is captured in the CCD camera and then analyzed in the PC connected to it. This arrangement can be used for tool wear compensation with the help of appropriate offline tool wear compensation algorithms (Yan et al., 2009; Malayath et al., 2019).

2.8 Tool Electrodes

Material for tool electrode in EDM should possess the essential qualities such as high MRR of workpiece and low tool wear rate. It is difficult to get all these properties in a tool electrode material due to its physical characteristics and properties and the conditions that occur in the sparking gap. In micro EDM, the material is also eroded out from the tool electrode due to melting and vaporization. The tool erosion or tool wear affects the shape and dimensional accuracy of the microfeatures. So, reducing tool wear has to be the main consideration while selecting a tool electrode for micro EDM.

The main considerations for selecting a tool electrode are as follows:

1. High thermal conductivity: reduces the electrode temperature
2. High melting and boiling point
3. High electrical conductivity
4. High machinability at low cost: to ease the tool preparation by conventional methods
5. High hardness: to withstand loads during a short circuit
6. Small grain size: increases surface quality.

Popular material choices for micro EDM tool electrodes are tungsten (W), tungsten carbide (WC), copper, graphite, brass, etc. Metallic electrodes have the advantages of high electrical conductivity and structural integrity when used in EDM. Properties of common materials used for micro EDM tool electrodes are given in Table 2.7.

Flexural and compressive strengths are important properties of tool electrodes. Many times, electrodes are broken or damaged due to rough handling or accidents. Higher strength also means that it is possible to machine tiny free-standing ribs and other fine details. Use of low-strength materials results in breaking and chipping during machining of thin walls and fine

TABLE 2.7

Common Electrode Materials for Micro EDM and Their Characteristics

Material	Wear Ratio	MRR	Fabrication	Cost
Copper	Low	High on a rough range	Easy	High
Brass	High	High only on finishing	Easy	Low
Tungsten	Lowest	Low	Difficult	High
Tungsten copper alloy	Low	Low	Difficult	High
Cast iron	Low	Low	Easy	Low
Steel	High	Low	Easy	Low
Zinc-based alloy	High	High on a rough range	Easy	High
Copper graphite	Low	High	Difficult	High

detail. Materials with flexural strengths above 10,000 psi (~69 MPa) are suitable for detailed working on the tool electrode. Thin-ribbed electrodes can be deflected by flushing pressure if the material does not have adequate flexural strength. Materials with high density along with tightly packed small particles will give better wear and finer finishes than materials with high density and large, loosely packed particles. The hardness of the material is particularly important when fabricating the electrode. Hard materials are difficult to machine without chipping the electrode. If the electrical resistivity is too high, tool electrode could be overheated while working with thin ribs and rods. This is because the energy would dissipate in the electrode, causing it to act like a heater. Electrical resistivity should be uniform for drilling of small hole with multiple rods (Poco Graphite, Inc.—www.edmtechman. com/about.cfm?pg=2&chap=5#a3).

Brass was one of the popularly used machining electrodes during the period of EDM development. However, the high wear rate of brass electrodes reduced the usability in micro EDM. Copper with appropriate pulse generator circuits and machining conditions proved to be usable for EDM machining with low wear rate. Copper and brass can remove materials at a high rate. However, they experience severe wear during discharge machining due to their relatively low melting points (1,084°C for copper and 930°C for brass yellow). The electrical current tends to concentrate at sharp edges and corners. The low melting point of copper limits its use in high current density. Furthermore, copper is more difficult to machine. It sticks to the cutter, clogs the grinding wheel, and produces burrs. So, for micro EDM, copper is not a right choice.

Tungsten has a high melting point (3,400°C) which reduces tool wear and increases MRR. However, the high hardness of tungsten makes it difficult to process or shape as required for making different features on the workpiece surface. Among the metallic electrodes, tungsten and its alloys are the popular choices for micro EDM tools due to their extraordinary stiffness and very low wear rate. Tungsten can be made more attractive for EDM by combining it with a more ductile material such as copper. The resulting material (tungsten copper) has many advantages such as high strength and hardness, good electrical and thermal conductivity, low thermal expansion coefficient, good arc resistance, high-temperature oxidation resistance, and good wear resistance. However, it is costlier than pure tungsten and copper.

Graphite has a melting point almost equal to that of tungsten. So, it has the advantages of high MRR and low tool wear rate. Machining of graphite is very easy as compared to copper and brass. Another advantage of graphite is that it does not produce any burrs during electrode fabrication. So, after machining of graphite to the desired shape, it can be used directly for EDM without any need of deburring operation. However, machining of graphite creates lots of dust, and it is a very serious problem for the machine, its operator as well as surrounding environment. Furthermore, it is difficult to produce fine features or small diameter due to low mechanical strength. Thin graphite electrodes are susceptible to breakage. Finish on the graphite surface depends on its particle size, microstructural consistency, and inherent physical properties. Electrodes made from high-strength Angstrofine (less than 1 μm particle size) and Ultrafine (1–5 μm particle size) graphite can tolerate much more pressure / force without chipping or breaking. The properties of graphite including density, thermal conductivity, bending strength, and electrical conductivity could be further improved by infiltrating melted metals into the porosities inside the graphite (e.g. copper graphite).

2.8.1 Special Tool Electrodes

To reduce the electrode wear during micro EDM, some special materials are used as tool electrode (Uhlmann and Roehner, 2008). This includes boron-doped chemical vapor deposition (CVD)-diamond (B-CVD) and polycrystalline diamond. To understand the potential of these materials, they are compared with other tool electrode materials in Figure 2.38. B-CVD exhibits less machining time and tool wear in micro EDM. The effect of grain size of B-CVD coatings are investigated (Uhlmann et al., 2010), and it is found that nanocrystalline coatings produced the smallest craters during micro EDM. To make the handling of microelectrodes easier, a peeling tool for micro EDM is developed (Tanabe et al., 2011). In this, the tool electrode is coated with a low-melting-point alloy, and the coating is peeled off during machining to expose the core electrode material (Tanabe et al., 2016). To machine high-aspect-ratio holes with minimum dimensional deviations, insulated tools can be a viable choice (Ferraris et al., 2013). Parylene C- and SiCN-SiC-coated tool electrodes outperformed their uncoated counterparts in deep micro EDM drilling, and an aspect ratio up to 126 can be achieved. Ti(C, N)-based cermets are also used in EDM as tool electrodes with improved MRR and reduced tool wear (Yoo et al., 2014). It is reported that

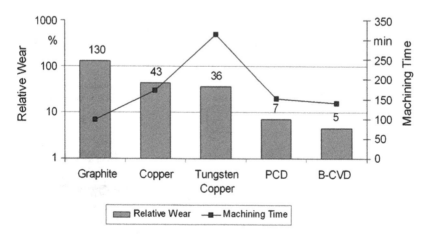

FIGURE 2.38
Influence of tool electrode materials on tool wear and machining time (Uhlmann and Roehner, 2008).

Ti-based, solid-solution carbonitrides electrode can reduce the tool wear rate by 35.38% and increase machining rate by 45.16% at optimal machining conditions. Micro EDM of single crystal SiC with the sintered diamond tool can remove the recast layer completely and fabricate a high-quality nanofinished machining surface (R_a = 1.85 nm) (Yan and Tan, 2015).

Laminated object manufacturing is used to fabricate 3D microelectrodes for EDM by superimposing 2D microstructures (Xu et al., 2015). The procedure of laminated electrode fabrication is shown in Figure 2.39. To reduce the effect of tool wear on the shape accuracy of machined features, a queue electrode system is developed in which one electrode does the roughing job followed by machining with finishing electrodes. Three-dimensional electrodes fabricated by localized electrochemical deposition (LECD) are also used in micro EDM tools (Habib and Rahman, 2016). To increase MRR during micro EDM, a foil-type tool electrode is used for microgrove machining (Yeo and Murali, 2003). Using foil-type electrode for microslitting has the advantages of less machining time and high machined surface quality. To push the limit of minimum achievable size in micro EDM, electrode materials with high electrical resistance are used (Koyano et al., 2017). Single crystal silicon is used as the electrode material, and it is found that the diameter of craters is reduced. Sheet-cylinder electrodes are used to drill microholes in ZrB$_2$-SiC-graphite composite (Li et al., 2017). The sheet segment of the electrode is used to drill the microhole with high capacitance followed by final machining with a cylinder part of the electrode. This electrode combination is helped to drill microholes in a composite material with a surface roughness (R_a) of 0.97 μm.

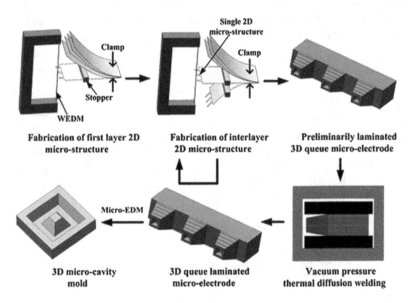

FIGURE 2.39
Fabrication of laminated electrodes for micro EDM (Xu et al., 2015).

Multifilament carbon fibers find their application as a potential complimentary electrode material in micro EDM (Trych-Wildner and Wildner, 2017). Recently, selective laser sintering has emerged as a fabrication method for complex tool electrodes for micro EDM (Uhlmann et al., 2018).

The possibility of avoiding the replication of the step effect of 3D microelectrodes in microfeatures was studied by Xu et al. (2018) with a method based on skin effect and point discharge. A liquid-phase alloy (Galinstan) has emerged as an interesting choice for micro EDM tool electrode recently (Huang et al., 2018). Yaou et al. (2017) developed a novel method in which electrostatic field-induced electrolyte jet is used as EDM tool electrode. To enhance flushing and debris removal during deep micro EDM drilling, using helical tools is an interesting solution (Plaza et al., 2014). Continuous removal of debris from the machining zone increases the machining efficiency and reduces machining time. It is reported that the helical tools with 45° helix angle and 50 µm flute length can reduce the machining time by 37% (Plaza et al., 2014). For a deeper hole with a flute length of 150 µm, the machining time is reduced by 19%. Hsue and Chang (2016) used tapered helical microtools made of tungsten carbide for grinding-assisted micro EDM. The tool is codeposited with diamond abrasives in a nickel–cobalt bath. Feasibility of using a tool electrode with an inclined microslit is studied by Kumar and Singh (2018).

2.9 Workpiece Materials

As micro EDM is employed for fabricating microparts related to various applications, the work material also changes accordingly. Hardness, electrical conductivity, and grain size of workpiece material are the factors that may affect the quality of the machined surface. A survey of different materials used in micro EDM is consolidated in Table 2.8.

2.10 Additional Technologies

EDG attachments: For in situ tool electrode fabrication, EDG systems are attached to the micro EDM machine tool. This includes a sacrificial block, disc, or wire-based tool arrangement (Asad et al., 2007).

PWM amplifier: A three-level PWM amplifier is developed to control the characteristics of magnetic suspension used for micro EDM (Guo et al., 2016). The amplifier circuit can reduce the radial and axial current ripples.

Workpiece actuation: A centrifugal servomechanism is developed to apply concentric roughing and eccentric finishing strategies in micro EDM (Amouzegar et al., 2016).

TABLE 2.8

Different Workpiece Materials Used in Micro EDM Experiments

Workpiece Material	Authors
Silicon wafer, silicon carbide, silicon nitride	Song et al. (2000), Muttamara et al. (2003, 2010) Yoo et al. (2015), Kumar Saxena et al. (2016), Saxena et al. (2016a), Liu and Huang (2000), Jahan et al. (2011), and Zhang et al. (2018)
Ti-6Al-4V	Murali and Yeo (2004), Pradhan et al. (2009), Kumar et al. (2012), Moses and Jahan (2015), and Feng et al. (2018)
Steel	Tai et al. (2007), Shabgard et al. (2009), Jahan et al. (2010b), D'Urso et al. (2014), Wang et al. (2017), Natarajan et al. (2013), and Kadirvel et al. (2013)
Diamond	Nakaoku et al. (2007) and Wang et al. (2011)
WC-Co	Song et al. (2010), Jahan et al. (2008, 2009a,b,c,e, 2010a,b,c), D'Urso et al. (2014), Hyun-Seok et al. (2009), and Wang et al. (2005)
Zirconia composites	Schubert et al. (2013), Huang and Yan (2015), Li et al. (2017), Schubert et al. (2011), Yeo et al. (2009c), and Ferraris et al. (2011)
Al and Al composites	Dave et al. (2014), Ferraris et al. (2011), and Tak et al. (2011)
Ni/Ti-Ni alloy	Perveen and Jahan (2018), Sing et al. (2018), Rasheed and Abidi (2012), Abidi et al. (2017), and Liu et al. (2005)
Carbon fiber-reinforced plastics	Teicher et al. (2013) and Kumar et al. (2018a)
Other materials	Wan et al. (2008)- *Polymethylmethacrylate/carbon nanotube composite* Singh (2010)- *Pyrolytic carbon* Dong et al. (2016)- *Be-Cu alloy* Li et al. (2009)- *TC4 alloy* Yan et al. (2002)- *Borosilicate glass*

Ultrasonic vibration: To increase the machining quality of EDM, vibration-assisted machining techniques are widely used. An ultrasonic resonant setup including pulse source, transducer, and horn or piezoelectric systems is used for exerting controlled vibrations in the system (Kumar et al., 2018b; Maity and Choubey, 2018). The debris particles are efficiently removed with the assistance of vibration in micro EDM.

2.11 Conclusions

One of the main aspects that differentiates micro EDM from conventional EDM is the level of sophistication it demands in terms of supporting technologies. The positioning systems must have high resolution, straightness, and good feedback control. The machine tool elements must have high vibrational damping capabilities. Special drive motors like linear drives have to be used to provide quick response to avoid machining discrepancies. The pulse generators must be capable of generating very short pulses with uniform discharge energy. Modified RC and transistor circuits are used to overcome the inherent limitations of the conventional pulse-generating circuits. An adaptive control strategy has to be employed to reduce the number of harmful discharges. Apart from commonly used dielectric fluids, special fluids and powder additives can be used to improve the machining characteristics. Special materials used for tool and workpiece in micro EDM are also discussed in detail. Conclusively, this chapter provides a comprehensive knowledge about the components and subsystems of micro EDM machine tool.

References

Abbas, N.M., Kunieda, M., 2016. Increasing discharge energy of micro-EDM with electrostatic induction feeding method through resonance in circuit. *Precis. Eng.* 45, 118–125. doi: 10.1016/j.precisioneng.2016.02.002.

Abidi, M.H., Al-ahmari, A.M., Siddiquee, A.N., 2017. An investigation of the micro-electrical discharge machining of nickel-titanium shape memory alloy using grey relations coupled with principal. *Metals* 2, 1–15. doi: 10.3390/met7110486.

Al-Ahmari, A.M.A., Rasheed, M.S., Mohammed, M.K., Saleh, T., 2016. A hybrid machining process combining micro-EDM and laser beam machining of nickel-titanium-based shape memory alloy. *Mater. Manuf. Process.* 31, 447–455. doi: 10.1080/10426914.2015.1019102.

Albinski, K., Musiol, K., Miernikiewicz, A., Labuz, S., Malota, M., 1996. The temperature of a plasma used in electrical discharge machining. *Plasma Sources Sci. Technol.* 5, 736.

Ali, M.Y., Atiqah, N., Erniyati, 2011. Silicon carbide powder mixed micro electro discharge milling of titanium alloy. *Int. J. Mech. Mater. Eng.* 6, 338–342. doi: 10.4028/www.scientific.net/AMR.383-390.1759.

Aligiri, E., Yeo, S.H., Tan, P.C., 2010. A new tool wear compensation method based on real-time estimation of material removal volume in micro-EDM. *J. Mater. Process. Technol.* 210, 2292–2303. doi: 10.1016/j.jmatprotec.2010.08.024.

Allaoua, B., Gasbaoui, B., Mebarki, B., 2009. Setting up PID DC motor speed control alteration parameters using particle swarm optimization strategy. *Leonardo Electron. J. Pract. Technol.* 7, 19–32. doi: 10.2174/9781608051267112010010003.

Allen, P., Chen, X., 2007. Process simulation of micro electro-discharge machining on molybdenum. *J. Mater. Process. Technol.* 186, 346–355. doi: 10.1016/j.jmatprotec.2007.01.009.

Altintas, Y., Verl, A., Brecher, C., Uriarte, L., Pritschow, G., 2011. Machine tool feed drives. *CIRP Ann. Manuf. Technol.* 60, 779–796. doi: 10.1016/j.cirp.2011.05.010.

Amouzegar, E., Ahmadi, S.R., Donyavi, A., 2016. State of the art micro-EDM drilling using centrifugal servo-mechanism for workpiece actuation. *Int. J. Adv. Manuf. Technol.* 85, 2027–2035. doi: 10.1007/s00170-015-7330-9.

Andromeda, T., Yahya, A., Khamis, N.H.H., Baharom, A., Rahim, M.A.A., 2012a. PID controller tuning by particle swarm optimization on electrical discharge machining servo control system. *ICIAS 2012-2012 4th Int. Conf. Intell. Adv. Syst. A Conf. World Eng. Sci. Technol. Congr. Conf. Proc.* 1, 51–55. doi: 10.1109/ICIAS.2012.6306157.

Andromeda, T., Yahya, A., Samion, S., Baharom, A., Hashim, N.L., 2012b. Differential evolution for optimization of pid gain in electrical discharge machining control system. *Trans. Can. Soc. Mech. Eng.* 37, 293–301.

Asad, A.B.M.A., Masaki, T., Rahman, M., Lim, H.S., Wong, Y.S., 2007. Tool-based micro-machining. *J. Mater. Process. Technol.* 192–193, 204–211. doi: 10.1016/j.jmatprotec.2007.04.038.

Ay, M., Çaydaş, U., Hasçalik, A., 2013. Optimization of micro-EDM drilling of inconel 718 superalloy. *Int. J. Adv. Manuf. Technol.* 66, 1015–1023. doi: 10.1007/s00170-012-4385-8.

Azad, M.S., Puri, A.B., 2012. Simultaneous optimisation of multiple performance characteristics in micro-EDM drilling of titanium alloy. *Int. J. Adv. Manuf. Technol.* 61, 1231–1239. doi: 10.1007/s00170-012-4099-y.

Behrens, A., Ginzel, J., 2003. Neuro-fuzzy process control system for sinking EDM. *J. Manuf. Process.* 5, 33–39. doi: 10.1016/S1526-6125(03)70038-3.

Bigot, S., D'Urso, G., Pernot, J.P., Merla, C., Surleraux, A., 2016. Estimating the energy repartition in micro electrical discharge machining. *Precis. Eng.* 43, 479–485. doi: 10.1016/j.precisioneng.2015.09.015.

Bissacco, G., Valentincic, J., Hansen, H.N., Wiwe, B.D., 2010. Towards the effective tool wear control in micro-EDM milling. *Int. J. Adv. Manuf. Technol.* 47, 3–9. doi: 10.1007/s00170-009-2057-0.

Blatnik, O., Valentincic, J., Junkar, M., 2007. Percentage of harmful discharges for surface current density monitoring in electrical discharge machining process. *Proc. Inst. Mech. Eng. Part B J. Eng. Manuf.* 221, 1677–1684. doi: 10.1243/09544054JEM816.

Boccadoro, M., Dauw, D.F., 1995. About the application of Fuzzy controllers in high-performance die-sinking EDM machines. *CIRP Ann. Manuf. Technol.* 44, 147–150. doi: 10.1016/S0007-8506(07)62294-X.

Byiringiro, J.B., Ikua, B.W., Nyakoe, G.N., 2009. Fuzzy logic based controller for micro-electro discharge machining servo systems. *IEEE AFRICON Conference*, Nairobi, Kenyab, 1–6. doi: 10.1109/AFRCON.2009.5308293.

Cao, H., Zhang, X., Chen, X., 2017. The concept and progress of intelligent spindles: A review. *Int. J. Mach. Tools Manuf.* 112, 21–52. doi: 10.1016/j.ijmachtools.2016.10.005.

Chang, Y.F., 2005. Mixed H_2/H_∞ optimization approach to gap control on EDM. *Control Eng. Pract.* 13, 95–104. doi: 10.1016/j.conengprac.2004.02.007.

Chang, Y.F., 2007. Robust PI controller design for EDM. *IECON 2007 - 33rd Annual Conference of the IEEE Industrial Electronics Society*, Taipei, Taiwan, 651–658. doi: 10.1109/IECON.2007.4459889.

Chen, S.T., Lai, Y.C., Liu, C.C., 2008. Fabrication of a miniature diamond grinding tool using a hybrid process of micro-EDM and co-deposition. *J. Micromech. Microeng.* 18, 055005. doi: 10.1088/0960-1317/18/5/055005.

Chen, T., Chen, Y., Hung, M., Hung, J., 2016. Design analysis of machine tool structure with artificial granite material. *Adv. Mech. Eng.* 8, 1–14. doi: 10.1177/1687814016656533.

Cheong, H., Kim, Y., Chu, C., 2018. Machining characteristics of an RC-type generator circuit with an N-channel MOSFET in micro EDM. *Procedia CIRP* 68, 631–636. doi: 10.1016/j.procir.2017.12.127.

Chi, G., Wang, Z., Xiao, K., Cui, J., Jin, B., 2008. The fabrication of a micro-spiral structure using EDM deposition in the air. *J. Micromech. Microeng.* 18, 1–9. doi: 10.1088/0960-1317/18/3/035027.

Chow, H.M., Yang, L.D., Lin, C.T., Chen, Y.F., 2008. The use of SiC powder in water as dielectric for micro-slit EDM machining. *J. Mater. Process. Technol.* 195, 160–170. doi: 10.1016/j.jmatprotec.2007.04.130.

Chung, C., Lu, S., Deng, J.S., 2009. Modeling and control of die-sinking EDM. *Proceedings of WSEAS International Conference on Mathematics and Computers in Science and Engineering*, World Scientific and Engineering Academy and Society.

Chung, D.K., Shin, H.S., Park, M.S., Chu, C.N., 2011. Machining characteristics of micro EDM in water using high frequency bipolar pulse. *Int. J. Precis. Eng. Manuf.* 12, 195–201. doi: 10.1007/s12541-011-0027-6.

Cogun, C., 1988. Computer aided evaluation and control of electric discharge machining by using properties of voltage pulse trains. *Proc. Res. Technol. Dev. Non-Tradit. Mach.* 134, 237–247.

Coğun, C., Savsar, M., 1990. Statistical modeling of properties discharge pulses in electric discharge machining. *Int. J. Mach. Tools Manuf.* 30, 467–474.

Cyril, J., Paravasu, A., Jerald, J., Sumit, K., Kanagaraj, G., 2017. Experimental investigation on performance of additive mixed dielectric during micro-electric discharge drilling on 316L stainless steel. *Mater. Manuf. Process.* 32, 638–644. doi: 10.1080/10426914.2016.1221107.

Dauw, D.F., Snoeys, R., Dekeyser, W., 1983. EDM electrode wear analysis by real time pulse detection. *Proceedings of 11th NAMRC*, Madison, WI, 372–378.

Dave, H.K., Mathai, V.J., Desai, K.P., Raval, H.K., 2014. Studies on quality of micro-holes generated on Al 1100 using micro-electro-discharge machining process. *Int. J. Adv. Manuf. Technol.* 76, 127–140. doi: 10.1007/s00170-013-5542-4.

Dong, S., Wang, Z., Wang, Y., Liu, H., 2016. An experimental investigation of enhancement surface quality of micro-holes for Be-Cu alloys using micro-EDM with multi-diameter electrode and different dielectrics. *Procedia CIRP*. doi: 10.1016/j.procir.2016.02.282.

Dunn, M.J., Clayton, F., Badger Meter Inc, 1964. Rotary drive electrical counting impulse generator. U.S. Patent 3,118,075.

D'Urso, G., Maccarini, G., Ravasio, C., 2014. Process performance of micro-EDM drilling of stainless steel. *Int. J. Adv. Manuf. Technol.* 72, 1287–1298. doi: 10.1007/s00170-014-5739-1.

Elsiti, N.M., Noordin, M.Y., 2017. Experimental investigations into the effect of process parameters and nano-powder (Fe$_2$O$_3$) on material removal rate during micro-EDM of Co-Cr-Mo. *Key Eng. Mater.* 740, 125–132. doi: 10.4028/www.scientific.net/KEM.740.125.

Feng, W., Chu, X., Hong, Y., Deng, D., 2017. Surface morphology analysis using fractal theory in micro electrical discharge machining. *Mater. Trans.* 58, 433–441. doi: 10.2320/matertrans.M2016381.

Feng, W., Chu, X., Hong, Y., Zhang, L., 2018. Studies on the surface of high-performance alloys machined by micro-EDM. *Mater. Manuf. Process.* 33, 616–625. doi: 10.1080/10426914.2017.1364758.

Fenggou, C., Dayong, Y., 2004. The study of high efficiency and intelligent optimization system in EDM sinking process. *J. Mater. Process. Technol.* 149, 83–87. doi: 10.1016/j.jmatprotec.2003.10.059.

Ferraris, E., Castiglioni, V., Ceyssens, F., Annoni, M., Lauwers, B., Reynaerts, D., 2013. EDM drilling of ultra-high aspect ratio micro holes with insulated tools. *CIRP Ann. Manuf. Technol.* 62, 191–194. doi: 10.1016/j.cirp.2013.03.115.

Ferraris, E., Reynaerts, D., Lauwers, B., 2011. Micro-EDM process investigation and comparison performance of Al 3O₂ and ZrO₂ based ceramic composites. *CIRP Ann. Manuf. Technol.* 60, 235–238. doi: 10.1016/j.cirp.2011.03.131.

Fu, X., Zhang, Q., Gao, L., Liu, Q., Wang, K., Zhang, Y.W., 2016. A novel micro-EDM—piezoelectric self-adaptive micro-EDM. *Int. J. Adv. Manuf. Technol.* 85, 817–824. doi: 10.1007/s00170-015-7939-8.

Fu, X.Z., Zhang, Y., Zhang, Q.H., Zhang, J.H., 2013. Research on piezoelectric self-adaptive Micro-EDM. *Procedia CIRP* 6, 303–308. doi: 10.1016/j.procir.2013.03.034.

Ghosh, A., Mallik, A.K., 1985. *Manufacturing Science.* East-West Press Private Limited, New Delhi.

Giandomenico, N., Gorgerat, F., Lavazais, B., 2016. Development of a new generator for Die Sinking electrical discharge machining. *Procedia CIRP* 42, 721–726. doi: 10.1016/j.procir.2016.02.308.

Giandomenico, N., Richard, J., Gorgerat, F., 2018. Smart generator for micro-EDM-milling. *Procedia CIRP* 68, 813–818. doi: 10.1016/j.procir.2017.12.161.

Guo, J., Zhang, G., Huang, Y., Ming, W., Liu, M., Huang, H., 2014. Investigation of the removing process of cathode material in micro-EDM using an atomistic-continuum model. *Appl. Surf. Sci.* 315, 323–336. doi: 10.1016/j.apsusc.2014.07.130.

Guo, R., Zhao, W., Li, G., 2006. Development of a cnc micro-edm machine. *International Technology and Innovation Conference 2006*, Hangzhou, China, 705–709. doi: 10.1049/cp:20060850.

Guo, Y., Ling, Z., 2016. A magnetic suspension spindle system for micro EDM. *Procedia CIRP* 42, 543–546. doi: 10.1016/j.procir.2016.02.248.

Guo, Y., Ling, Z., Zhang, X., 2016. A novel PWM power amplifier of magnetic suspension spindle control system for micro EDM. *Int. J. Adv. Manuf. Technol.* 83, 961–973. doi: 10.1007/s00170-015-7622-0.

Guo, Y., Ling, Z., Zhang, X., Feng, Y., 2018. A magnetic suspension spindle system for small and micro holes EDM. *Int. J. Adv. Manuf. Technol.* 94, 1911–1923. doi: 10.1007/s00170-017-0990-x.

Habib, M.A., Rahman, M., 2016. Performance of electrodes fabricated by localized electrochemical deposition (LECD) in micro-EDM operation on different workpiece materials. *J. Manuf. Process.* 24, 78–89. doi: 10.1016/j.jmapro.2016.08.003.

Han, F., Wachi, S., Kunieda, M., 2004. Improvement of machining characteristics of micro-EDM using transistor type isopulse generator and servo feed control. *Precis. Eng.* 28, 378–385.

Hao, T., Yang, W., Yong, L., 2008. Vibration-assisted servo scanning 3D micro EDM. *J. Micromech. Microeng.* 18. doi: 10.1088/0960-1317/18/2/025011.

Hara, S., Nishioki, N., 2002. Ultra-high speed discharge control for micro electric discharge machining, in: I. Inasak, ed. *Initiatives of Precision Engineering at the Beginning of a Millennium.* Springer, Berlin, pp. 194–198.

Hashim, N.L.S., Yahya, A., Daud, M.R., Syahrullail, S., Baharom, A., Khamis, N.H., Mahmud, N., 2015. Review on an electrical discharge machining servomechanism system. *Sci. Iran.* 22, 1813–1832.

He, D., Morita, H., Zhang, X., Shinshi, T., Nakagawa, T., Sato, T., Miyake, H., 2010. Development of a novel 5-DOF controlled maglev local actuator for high-speed electrical discharge machining. *Precis. Eng.* 34, 453–460. doi: 10.1002/ejoc.201500973.

Hebbar, R.R., Ramabhadran, R., Chandrasekar, S., Cummins Inc, 2002. Hybrid servomechanism for micro-electrical discharge machining. U.S. Patent 6,385,500.

Heinz, K., Kapoor, S.G., DeVor, R.E., Surla, V., 2011. An investigation of magnetic-field-assisted material removal in micro-EDM for nonmagnetic materials. *J. Manuf. Sci. Eng.* 133, 021002. doi: 10.1115/1.4003488.

Heo, S., Jeong, Y.H., Min, B.K., Lee, S.J., 2009. Virtual EDM simulator: Three-dimensional geometric simulation of micro-EDM milling processes. *Int. J. Mach. Tools Manuf.* 49, 1029–1034. doi: 10.1016/j.ijmachtools.2009.07.005.

Herzig, M., Berger, T., Schulze, H.P., Hackert-Oschätzchen, M., Kröning, O., Schubert, A., 2017. Modification of the process dynamics in micro-EDM by means of an additional piezo-control system. *AIP Conf. Proc.* 1896. doi: 10.1063/1.5008048.

Higuchi, T., Furutani, K., Yamagata, Y., Takeda, K., Makino, H., 1991. Development of pocket-size electro-discharge machine. *CIRP Ann. Manuf. Technol.* 40, 203–206. doi: 10.1016/S0007-8506(07)61968-4.

Hongzhe, Z., Shusheng, B., Bo, P., 2015. Dynamic analysis and experiment of a novel ultra-precision compliant linear-motion mechanism. *Precis. Eng.* 42, 352–359. doi: 10.1016/j.precisioneng.2015.06.002.

Hsieh, M.F., Tung, C.J., Yao, W.S., Wu, M.C., Liao, Y.S., 2007. Servo design of a vertical axis drive using dual linear motors for high speed electric discharge machining. *Int. J. Mach. Tools Manuf.* 47, 546–554. doi: 10.1016/j.ijmachtools.2006.05.011.

Hsue, A.W., Chung, C., 2009. Control strategy for high speed electrical discharge machining (die-sinking EDM) equipped with linear motors. *International Conference on Advanced Intelligent Mechatronics*, Singapore, 326–331.

Hsue, A.W.J., Chang, Y.F., 2016. Toward synchronous hybrid micro-EDM grinding of micro-holes using helical taper tools formed by Ni-Co/diamond Co-deposition. *J. Mater. Process. Technol.* 234, 368–382. doi: 10.1016/j.jmatprotec.2016.04.009.

Hu, M., Zhou, Z., Li, Y., Du, H., 2001. Development of a linear electrostrictive servo motor. *Precis. Eng.* 25, 316–320. doi: 10.1016/S0141-6359(01)00085-X.

Hua, H., Yang, F., Yang, J., Cao, Y., Li, C., Peng, F., 2018. Reanalysis of discharge voltage of RC-type generator in micro-EDM. *Procedia CIRP* 68, 625–630. doi: 10.1016/j.procir.2017.12.126.

Huang, H., Yan, J., 2015. On the surface characteristics of a Zr-based bulk metallic glass processed by microelectrical discharge machining. *Appl. Surf. Sci.* 355, 1306–1315. doi: 10.1016/j.apsusc.2015.08.239.

Huang, R., Yi, Y., Yu, W., Takahata, K., 2018. Liquid-phase alloy as a microfluidic electrode for micro-electro-discharge patterning. *J. Mater. Process. Technol.* 258, 1–8. doi: 10.1016/j.jmatprotec.2018.03.012.

Hung, J.C., Yang, T.C., Li, K.C., 2011. Studies on the fabrication of metallic bipolar plates - using micro electrical discharge machining milling. *J. Power Sources* 196, 2070–2074. doi: 10.1016/j.jpowsour.2010.10.001.

Huo, D., 2013. *Micro-Cutting: Fundamentals and Applications*. John Wiley & Sons, Chichester.

Hyun-Seok, T.A.K., Chang-Seung, H.A., Dong-Hyun, K.I.M., Ho-Jun, L.E.E., Hae-June, L.E.E., Myung-Chang, K., 2009. Comparative study on discharge conditions in micro-hole electrical discharge machining of tungsten carbide (WC-Co) material. *Trans. Nonferrous Met. Soc. China* 19, S114–S118.

Imai, Y., Nakagawa, T., Miyake, H., Hidai, H., Tokura, H., 2004. Local actuator module for highly accurate micro-EDM. *J. Mater. Process. Technol.* 149, 328–333. doi: 10.1016/j.jmatprotec.2004.01.060.

Imai, Y., Satake, A., Taneda, A., Kobayashi, K., 1996. Improvement of EDM machining speed by using high frequency response actuator. *Int. J. Electr. Mach.* 30, 21–26.

Jaen-Cuellar, A.Y., Romero-Troncoso, R.D.J., Morales-Velazquez, L., Osornio-Rios, R.A., 2013. PID-controller tuning optimization with genetic algorithms in servo systems. *Int. J. Adv. Robot. Syst.* 10. doi: 10.5772/56697.

Jahan, M.P., Anwar, M.M., Wong, Y.S., Rahman, M., 2009a. Nanofinishing of hard materials using micro-electrodischarge machining. *Proc. Inst. Mech. Eng. Part B J. Eng. Manuf.* 223, 1127–1142.

Jahan, M.P., Lieh, T.W., Wong, Y.S., Rahman, M., 2011. An experimental investigation into the micro-electrodischarge machining behavior of p-type silicon. *Int. J. Adv. Manuf. Technol.* 57, 617–637. doi: 10.1007/s00170-011-3302-x.

Jahan, M.P., Rahman, M., Wong, Y.S., 2010a. Migration of materials during finishing micro-EDM of tungsten carbide. *Key Eng. Mater.* 443, 681–686.

Jahan, M.P., Rahman, M., Wong, Y.S., 2014. *Micro-Electrical Discharge Machining (Micro-EDM): Processes, Varieties, and Applications, Comprehensive Materials Processing.* Elsevier. doi: 10.1016/B978-0-08-096532-1.01107-9.

Jahan, M.P., Wong, Y.S., Rahman, M., 2008. A comparative study of transistor and RC pulse generators for micro-EDM of tungsten carbide. *Int. J. Precis. Eng. Manuf.* 9, 3–10.

Jahan, M.P., Wong, Y.S., Rahman, M., 2009b. A study on the quality micro-hole machining of tungsten carbide by micro-EDM process using transistor and RC-type pulse generator. *J. Mater. Process. Technol.* 209, 1706–1716. doi: 10.1016/j.jmatprotec.2008.04.029.

Jahan, M.P., Wong, Y.S., Rahman, M., 2009c. A study on the fine-finish die-sinking micro-EDM of tungsten carbide using different electrode materials. *J. Mater. Process. Technol.* 209, 3956–3967. doi: 10.1016/j.jmatprotec.2008.09.015.

Jahan, M.P., Wong, Y.S., Rahman, M., 2009d. Effect of non-electrical and gap control parameters in the micro-EDM of WC-Co. *J Mach Form Technol* 1, 51–78.

Jahan, M.P., Wong, Y.S., Rahman, M., 2010b. A comparative experimental investigation of deep-hole micro-EDM drilling capability for cemented carbide (WC-Co) against austenitic stainless steel (SUS 304). *Int. J. Adv. Manuf. Technol.* doi: 10.1007/s00170-009-2167-8.

Jahan, M.P., Wong, Y.S., Rahman, M., 2010c. A comparative study on the performance of sinking and milling micro-EDM for nanofinishing of tungsten carbide. *Int. J. Nanomanuf.* 6, 190–206. doi: 10.1504/IJNM.2010.034783.

Jahan, M.P., Wong, Y.S., Rahman, M., Asad, A., 2009e. An experimental investigation on the surface characteristics of tungsten carbide for the fine-finish die-sinking and scanning micro-EDM. *Int. J. Abras. Technol.* 2, 223–244.

Jain, T., Nigam, M.J., 2009. Optimization of PD-PI controller using Swarm. *Int. J.* 6, 55–59.

Jameson, E.C., 2001. Description and development of elcetrical discharge machining, in: E.C. Jameson, ed. *Electrical Discharge Machining.* Society of Manufacturing Engineers, Southfield, MI, p. 329.

Jeong, Y.H., HanYoo, B., Lee, H.U., Min, B.K., Cho, D.W., Lee, S.J., 2009. Deburring microfeatures using micro-EDM. *J. Mater. Process. Technol.* 209, 5399–5406. doi: 10.1016/j.jmatprotec.2009.04.021.

Jia, Z., Zhang, L., Wang, F., Liu, W., 2010. A new method for discharge state prediction of micro-EDM using empirical mode decomposition. *J. Manuf. Sci. Eng.* 132, 014501. doi: 10.1115/1.4000559.

Jiang, Y., Zhao, W., Xi, X., Gu, L., 2012a. Adaptive control for small-hole EDM process with wavelet transform detecting method. *J. Mech. Sci. Technol.* 26, 1885–1890. doi: 10.1007/s12206-012-0410-y.

Jiang, Y., Zhao, W., Xi, X., Gu, L., Kang, X., 2012b. Detecting discharge status of small-hole EDM based on wavelet transform. *Int. J. Adv. Manuf. Technol.* 61, 171–183. doi: 10.1007/s00170-011-3676-9.

Jung, J.W., Jeong, Y.H., Min, B.-K., Lee, S.J., 2008. Model-based pulse frequency control for micro-EDM milling using real-time discharge pulse monitoring. *J. Manuf. Sci. Eng.* 130, 031106. doi: 10.1115/1.2917305.

Kadirvel, A., Hariharan, P., Gowri, S., 2013. Experimental investigation on the electrode specific performance in micro-EDM of die-steel. *Mater. Manuf. Process.* 28, 390–396. doi: 10.1080/10426914.2013.763959.

Kaneko, T., Onodera, T., 2004. Improvement in machining performance of die-sinking EDM by using self-adjusting fuzzy control. *J. Mater. Process. Technol.* 149, 204–211. doi: 10.1016/j.jmatprotec.2004.02.006.

Kao, C.C., Shih, A.J., 2006. Sub-nanosecond monitoring of micro-hole electrical discharge machining pulses and modeling of discharge ringing. *Int. J. Mach. Tools Manuf.* 46, 1996–2008. doi: 10.1016/j.ijmachtools.2006.01.008.

Kao, C.C., Shih, A.J., Miller, S.F., 2008. Fuzzy logic control of microhole electrical discharge machining. *J. Manuf. Sci. Eng.* 130, 064502. doi: 10.1115/1.2977827.

Kibria, G., Sarkar, B.R., Pradhan, B.B., Bhattacharyya, B., 2010. Comparative study of different dielectrics for micro-EDM performance during microhole machining of Ti-6Al-4V alloy. *Int. J. Adv. Manuf. Technol.* 48, 557–570. doi: 10.1007/s00170-009-2298-y.

Kim, D.I., Jung, S.C., Lee, J.E., Chang, S.H., 2006a. Parametric study on design of composite-foam-resin concrete sandwich structures for precision machine tool structures. *Compos. Struct.* 75, 408–414. doi: 10.1016/j.compstruct.2006.04.022.

Kim, D.J., Yi, S.M., Lee, Y.S., Chu, C.N., 2006b. Straight hole micro EDM with a cylindrical tool using a variable capacitance method accompanied by ultrasonic vibration. *J. Micromech. Microeng.* 16, 1092–1097. doi: 10.1088/0960-1317/16/5/031.

Klocke, F., Zeis, M., Heidemanns, L., 2018. Fluid structure interaction of thin graphite electrodes during flushing movements in sinking electrical discharge machining. *CIRP J. Manuf. Sci. Technol.* 20, 23–28. doi: 10.1016/j.cirpj.2017.09.003.

Koyano, T., Sugata, Y., Hosokawa, A., Furumoto, T., 2017. Micro electrical discharge machining using high electric resistance electrodes. *Precis. Eng.* 47, 480–486. doi: 10.1016/j.precisioneng.2016.10.003.

Kröning, O., Herzig, M., Hackert-oschätzchen, M., Kühn, R., Zeidler, H., Schubert, A., 2015. Micro electrical discharge machining of tungsten carbide with ultra-short pulse 653, 759–764. doi: 10.4028/www.scientific.net/KEM.651-653.759.

Kruth, J.P., Snoeys, R., van Brussel, H., 1979a. In-process optimization of electro-discharge machining. *Proceedings of the Nineteenth International Machine Tool Design and Research Conference*, Springer, 567–574.

Kruth, J.P., Snoeys, R., van Brussel, H., 1979b. Adaptive control optimization of the EDM process using minicomputers. *Comput Ind.* 1(2), 65–75.

Kumar, P., Singh, P.K., Kumar, D., Prakash, V., Hussain, M., Das, A.K., 2017. A novel application of micro-EDM process for the generation of nickel nanoparticles with different shapes. *Mater. Manuf. Process.* 32, 564–572. doi: 10.1080/10426914.2016.1244832.

Kumar, R., Kumar, A., Singh, I., 2018a. Electric discharge drilling of micro holes in CFRP laminates. *J. Mater. Process. Technol.* 259, 150–158. doi: 10.1016/j.jmatprotec.2018.04.031.

Kumar, R., Singh, I., 2018. Productivity improvement of micro EDM process by improvised tool. *Precis. Eng.* 51, 529–535. doi: 10.1016/j.precisioneng.2017.10.008.

Kumar, S., Grover, S., Walia, R.S., 2018b. Analyzing and modeling the performance index of ultrasonic vibration assisted EDM using graph theory and matrix approach. *Int. J. Interact. Des. Manuf.* 12, 225–242. doi: 10.1007/s12008-016-0355-y.

Kumar, S., Singh, R., Batish, A., Singh, T.P., 2012. Electric discharge machining of titanium and its alloys: A review. *Int. J. Mach. Mach. Mater.* 11, 84–111. doi: 10.1504/IJMMM.2012.044922.

Kumar Saxena, K., Suman Srivastava, A., Agarwal, S., 2016. Experimental investigation into the micro-EDM characteristics of conductive SiC. *Ceram. Int.* 42, 1597–1610. doi: 10.1016/j.ceramint.2015.09.111.

Kumar, V., Rana, K.P.S., Gupta, V., 2008. Real-time performance evaluation of a Fuzzy PI + Fuzzy PD controller for liquid-level process. *Int. J. Intell. Control Syst.* 13, 89–96. doi: 10.1016/j.apal.2006.10.001.

Kunieda, M., Hayasaka, A., Yang, X.D., Sano, S., Araie, I., 2007. Study on nano EDM using capacity coupled pulse generator. *CIRP Ann. Manuf. Technol.* 56, 213–216. doi: 10.1016/j.cirp.2007.05.051.

Kunieda, M., Lauwers, B., Rajurkar, K.P., Schumacher, B.M., 2005. Advancing EDM through fundamental insight into the process. *CIRP Ann. Manuf. Technol.* 54, 64–87. doi: 10.1016/S0007-8506(07)60020-1.

Kuo, J.-L., Chang, Z.-S., Lee, J.-D., Chio, Y.-C., Lee, P.-H., 2002. Ringing effect analysis of the digital current pulse generator for the linear rail gun. *Conference Record of the IEEE Industry Applications Conference, 37th IAS Annual Meeting*, Pittsburgh, PA, 176–181.

Kurnia, W., Tan, P.C., Yeo, S.H., Wong, M., 2008. Analytical approximation of the erosion rate and electrode wear in micro electrical discharge machining. *J. Micromech. Microeng.* 18. doi: 10.1088/0960-1317/18/8/085011.

Lee, H.S., Tomizuka, M., 1996. Robust motion controller design for high-accuracy positioning systems. *IEEE Trans. Ind. Electron.* 43(1), 48–55.

Li, H., Wang, Z., Wang, Y., Liu, H., Bai, Y., 2017. Micro-EDM drilling of ZrB2-SiC-graphite composite using micro sheet-cylinder tool electrode. *Int. J. Adv. Manuf. Technol.* 92, 2033–2041. doi: 10.1007/s00170-017-0296-z.

Li, L., Diver, C., Atkinson, J., Giedl-Wagner, R., Helml, H.J., 2006. Sequential laser and EDM micro-drilling for next generation fuel injection nozzle manufacture. *CIRP Ann. Manuf. Technol.* 55, 179–182. doi: 10.1016/S0007-8506(07)60393-X

Li, M.S., Chi, G.X., Wang, Z.L., Wang, Y.K., Dai, L., 2009. Micro electrical discharge machining of small hole in TC4 alloy. *Trans. Nonferrous Met. Soc. China*, English Ed. 19, S434–S439. doi: 10.1016/S1003-6326(10)60084-2.

Li, Q., Bai, J., Li, C., Li, S., 2013. Research on multi-mode pulse power supply for array micro holes machining in micro-EDM. *Procedia CIRP* 6, 168–173. doi: 10.1016/j.procir.2013.03.088.

Li, Y., Guo, M., Zhou, Z., Hu, M., 2002. Micro electro discharge machine with an inchworm type of micro feed mechanism. *Precis. Eng.* 26, 7–14. doi: 10.1016/S0141-6359(01)00088-5.

Li, Y., Zhao, W., Lan, S., Ni, J., Wu, W., Lu, B., 2015. A review on spindle thermal error compensation in machine tools. *Int. J. Mach. Tools Manuf.* 95, 20–38. doi: 10.1016/j.ijmachtools.2015.04.008.

Liang, H.Y., Kuo, C.L., Huang, J.D., 2002. Precise micro-assembly through an integration of micro-EDM and Nd-YAG. *Int. J. Adv. Manuf. Technol.* 20, 454–458. doi: 10.1007/s001700200177.

Liao, Y.S., Chang, T.Y., Chuang, T.J., 2008. An on-line monitoring system for a micro electrical discharge machining (micro-EDM) process. *J. Micromech. Microeng.* 18. doi: 10.1088/0960-1317/18/3/035009.

Lin, C.S., Liao, Y.S., Cheng, Y.T., Lai, Y.C., 2010. Fabrication of micro ball joint by using micro-EDM and electroforming. *Microelectron. Eng.* 87, 1475–1478. doi: 10.1016/j. mee.2009.11.087.

Lin, M.Y., Tsao, C.C., Huang, H.H., Wu, C.Y., Hsu, C.Y., 2015. Use of the grey-Taguchi method to optimise the micro-electrical discharge machining (micro-EDM) of Ti-6Al-4V alloy. *Int. J. Comput. Integr. Manuf.* 28, 569–576. doi: 10.1080/0951192X.2014.880946.

Liu, C., Huang, J., 2000. Micro-electrode discharge machining of TiN/Si$_3$N$_4$ composites. *Br. Ceram. Trans.* 99, 149–152. doi: 10.1179/096797800680866.

Liu, H.S., Yan, B.H., Huang, F.Y., Qiu, K.H., 2005. A study on the characterization of high nickel alloy micro-holes using micro-EDM and their applications. *J. Mater. Process. Technol.* 169, 418–426. doi: 10.1016/j.jmatprotec.2005.04.084.

Liu, W., Jia, Z., Zou, S., Zhang, L., 2014. A real-time predictive control method of discharge state for micro-EDM based on calamities grey prediction theory. *Int. J. Adv. Manuf. Technol.* 72, 135–144. doi: 10.1007/s00170-014-5644-7.

Liu, Y., Zeng, W.L., 2011. Research on Fuzzy control algorithm with double parameters detection of micro-EDM. *Appl. Mech. Mater.* 65, 75–78. doi: 10.4028/www. scientific.net/AMM.65.75.

Maas, J., 2000. Model-based control for ultrasonic motors. *IEEE/ASME Trans. Mechatron.* 5, 165–180. doi: 10.1109/3516.847090.

Mahardika, M., Mitsui, K., 2008. A new method for monitoring micro-electric discharge machining processes. *Int. J. Mach. Tools Manuf.* 48, 446–458. doi: 10.1016/j. ijmachtools.2007.08.023.

Mahdavinejad, R.A., 2009. EDM process optimisation via predicting a controller model. *Arch. Comput. Mater. Sci. Surf. Eng.* 1, 161–167.

Mahmud, N., Yahya, A., Rafiq, M., Samion, S., Safura, N.L., 2012. Electrical discharge machining pulse power generator to machine micropits of hip implant. *2012 International Conference on Biomedical Engineering (ICoBE)*, Penang, Malaysia, 493–497. doi: 10.1109/ICoBE.2012.6179066.

Mahyar, S., Moghadam, M., Alibeiki, E., Khosravi, A., 2018. Adaptive control of machining process using Electrical Discharging Method (EDM) based on Self-Tuning Regulator (STR). *Int. J. of Smart Electr. Eng.* 7(1), 9–16.

Maity, K.P., Choubey, M., 2018. A review on vibration-assisted EDM, micro-EDM and WEDM. *Surf. Rev. Lett.* doi: 10.1142/S0218625X18300083.

Malayath, G., Katta, S., Sidpara, A.M., Deb, S., 2019. Length-wise tool wear compensation for micro electric discharge drilling of blind holes. *Measurement.* 134, 888–896.

Manivannan, R., Kumar, M.P., 2016. Multi-response optimization of micro-EDM process parameters on AISI304 steel using TOPSIS. *J. Mech. Sci. Technol.* 30, 137–144. doi: 10.1007/s12206-015-1217-4.

Manivannan, R., Kumar, M.P., 2018. Improving the machining performance characteristics of the µEDM drilling process by the online cryogenic cooling approach. *Mater. Manuf. Process.* 33, 390–396. doi: 10.1080/10426914.2017.1303145.

Masuzawa, T., 1975. Improvement of micro-EDM machining speed by using voice coil actuator. *J. JSEME* 8, 43–52.

Mathai, V.J., Dave, H.K., Desai, K.P., 2018. End wear compensation during planetary EDM of Ti–6Al–4V by adaptive neuro fuzzy inference system. *Prod. Eng.* 12, 1–10. doi: 10.1007/s11740-017-0778-8.

Mechanics, A., 2014. Development of a new pulse discriminating system for micro Electro Discharge Machining (µEDM). *Appl. Mech. Mater.* 542, 430–435. doi: 10.4028/www.scientific.net/AMM.541-542.430.

Mehfuz, R., Ali, M.Y., 2009. Investigation of machining parameters for the multiple-response optimization of micro electrodischarge milling. *Int. J. Adv. Manuf. Technol.* 43, 264–275. doi: 10.1007/s00170-008-1705-0.

Modica, F., Guadagno, G., Marrocco, V., Fassi, I., 2014. Evaluation of micro-EDM milling performance using pulse discrimination. *ASME 2014 International Design Engineering Technical Conferences and Computers and Information in Engineering Conference*, Buffalo. American Society of Mechanical Engineers, V004T09A019–V004T09A019.

Modica, F., Marrocco, V., Valori, M., Viganò, F., Annoni, M., Fassi, I., 2018. Study about the influence of powder mixed water based fluid on micro-EDM process. *Procedia CIRP* 68, 789–795. doi: 10.1016/j.procir.2017.12.156.

Mohammad, Y.A., 2012. Powder mixed micro electro discharge milling of titanium alloy : Analysis of surface roughness. *Adv. Mater. Res.* 342, 142–146. doi: 10.4028/www.scientific.net/AMR.341-342.142.

Möhring, H.C., Brecher, C., Abele, E., Fleischer, J., Bleicher, F., 2015. Materials in machine tool structures. *CIRP Ann. Manuf. Technol.* 64, 725–748. doi: 10.1016/j.cirp.2015.05.005.

Moses, M.D., Jahan, M.P., 2015. Micro-EDM machinability of difficult-to-cut Ti-6Al-4V against soft brass. *Int. J. Adv. Manuf. Technol.* 81, 1345–1361. doi: 10.1007/s00170-015-7306-9.

Moylan, S.P., Chandrasekar, S., Benavides, G.L., 2005. High-speed micro-electro-discharge machining. *Sandia Rep.* 61, 379–386.

Murali, M., Yeo, S.H., 2004. Rapid biocompatible micro device fabrication by micro electro-discharge machining. *Biomed. Microdevices* 6, 41–45. doi: 10.1023/B:BMMD.0000013364.71148.51.

Muralidhara, Vasa, N.J., Makaram, S., 2009. Investigations on a directly coupled pie-zoactuated tool feed system for micro-electro-discharge machine. *Int. J. Mach. Tools Manuf.* 49, 1197–1203. doi: 10.1016/j.ijmachtools.2009.08.004.

Muthuramalingam, T., Mohan, B., 2014. Performance analysis of iso current pulse generator on machining characteristics in EDM process. *Arch. Civ. Mech. Eng.* 14, 383–390. doi: 10.1016/j.acme.2013.10.003.

Muttamara, A., Fukuzawa, Y., Mohri, N., Tani, T., 2003. Probability of precision micro-machining of insulating Si_3N_4 ceramics by EDM. *J. Mater. Process. Technol.* 140, 243–247. doi: 10.1016/S0924-0136(03)00745-3.

Muttamara, A., Janmanee, P., Fukuzawa, Y., 2010. A study of micro – EDM on silicon nitride using electrode materials. *Int. Trans. J. Eng. Appl. Sci. Technol.* 1, 1–7.

Nakaoku, H., Masuzawa, T., Fujino, M., 2007. Micro-EDM of sintered diamond. *J. Mater. Process. Technol.* 187–188, 274–278. doi: 10.1016/j.jmatprotec.2006.11.082.

Narasimhan, J., Yu, Z., Rajurkar, K.P., 2005. Tool wear compensation and path generation in micro and macro EDM. *J. Manuf. Process.* 7, 75–82. doi: 10.1016/S1526-6125(05)70084-0.

Natarajan, N., Arunachalam, R.M., Thanigaivelan, R., 2013. Experimental study and analysis of micro holes machining in EDM of SS 304. *Int. J. Mach. Mater.* 13, 1. doi: 10.1504/IJMMM.2013.051905.

Natarajan, N., Suresh, P., 2015. Experimental investigations on the microhole machining of 304 stainless steel by micro-EDM process using RC-type pulse generator. *Int. J. Adv. Manuf. Technol.* 77, 1741–1750. doi: 10.1007/s00170-014-6494-z.

Nguyen, M.D., Wong, Y.S., Rahman, M., 2013. Profile error compensation in high precision 3D micro-EDM milling. *Precis. Eng.* 37, 399–407. doi: 10.1016/j.precisioneng.2012.11.002.

Nirala, C.K., Reddy, B., Saha, P., 2012. Optimization of process parameters in micro electro- discharge drilling [micro EDM-Drilling]: A taguchi approach. *Adv. Mater. Res.* 622–623, 30–34. doi: 10.4028/www.scientific.net/AMR.622-623.30.

Nirala, C.K., Saha, P., 2015. Development of an algorithm for online pulse discrimination in micro-EDM using current and voltage sensors and their comparison. *Souvenir 2015 IEEE International Advance Computing Conference (IACC)*, Bangalore, India, 496–500. doi: 10.1109/IADCC.2015.7154758.

Nirala, C.K., Saha, P., 2016. A new approach of tool wear monitoring and compensation in RμEDM process. *Mater. Manuf. Process.* 31, 483–494. doi: 10.1080/10426914.2015.1058950.

Nirala, C.K., Unune, D.R., Sankhla, H.K., 2017. Virtual signal-based pulse discrimination in micro-electro-discharge machining. *J. Manuf. Sci. Eng.* 139, 094501. doi: 10.1115/1.4037108.

Peng, Z.L., Wang, Z.L., Jin, B.D., 2008. Micro-forming process and microstructure of deposit by using micro EDM deposition in air. *Key Eng. Mater.* 375–376, 153–157. doi: 10.4028/www.scientific.net/KEM.375-376.153.

Perveen, A., Jahan, M.P., 2018. Comparative micro-EDM studies on Ni based X-alloy using coated and uncoated tools. *Mater. Sci. Forum* 911, 13–19. doi: 10.4028/www.scientific.net/MSF.911.13.

Perveen, A., Jahan, M.P., Zhumagulov, S., 2018. Statistical modeling and optimization of micro electro discharge machining of Ti alloy. *Mater. Today Proc.* 5, 4803–4810. doi: 10.1016/j.matpr.2017.12.054.

Pham, D.T., Dimov, S.S., Bigot, S., Ivanov, A., Popov, K., 2004. Micro-EDM - recent developments and research issues. *J. Mater. Process. Technol.* 149, 50–57. doi: 10.1016/j.jmatprotec.2004.02.008.

Pham, D.T., Ivanov, A., Bigot, S., Popov, K., Dimov, S., 2007. An investigation of tube and rod electrode wear in micro EDM drilling. *Int. J. Adv. Manuf. Technol.* 33, 103–109. doi: 10.1007/s00170-006-0639-7.

Plaza, S., Sanchez, J.A., Perez, E., Gil, R., Izquierdo, B., Ortega, N., Pombo, I., 2014. Experimental study on micro EDM-drilling of Ti6Al4V using helical electrode. *Precis. Eng.* 38, 821–827. doi: 10.1016/j.precisioneng.2014.04.010.

Pradhan, B.B., Masanta, M., Sarkar, B.R., Bhattacharyya, B., 2009. Investigation of electro-discharge micro-machining of titanium super alloy. *Int. J. Adv. Manuf. Technol.* 41, 1094–1106. doi: 10.1007/s00170-008-1561-y.

Prihandana, G.S., Mahardika, M., Hamdi, M., Mitsui, K., 2009a. Effect of m micro MoS2 powder mixed dielectric fluid on surface quality and material removal rate in micro EDM-processes. *Trans. Mater. Res. Soc. Jpn.* 34, 329–332. doi: 10.14723/tmrsj.34.329.

Prihandana, G.S., Mahardika, M., Hamdi, M., Wong, Y.S., Mitsui, K., 2009b. Effect of micro-powder suspension and ultrasonic vibration of dielectric fluid in micro-EDM processes-Taguchi approach. *Int. J. Mach. Tools Manuf.* 49, 1035–1041. doi: 10.1016/j.ijmachtools.2009.06.014.

Prihandana, G.S., Mahardika, M., Hamdi, M., Wong, Y.S., Mitsui, K., 2011. Accuracy improvement in nanographite powder-suspended dielectric fluid for micro-electrical discharge machining processes. *Int. J. Adv. Manuf. Technol.* 56, 143–149. doi: 10.1007/s00170-011-3152-6.

Prihandana, G.S., Sriani, T., Mahardika, M., Hamdi, M., Miki, N., Wong, Y.S., Mitsui, K., 2014. Application of powder suspended in dielectric fluid for fine finish micro-EDM of Inconel 718. *Int. J. Adv. Manuf. Technol.* 75, 599–613. doi: 10.1007/s00170-014-6145-4.

Qian, J., Yang, F., Wang, J., Lauwers, B., Reynaerts, D., 2015. Material removal mechanism in low-energy micro-EDM process. *CIRP Ann.* 64, 225–228.

Snoeys, R., Dauw, D., Kruth, J.P., 1980. Improved adaptive control system for. *CIRP Ann. Manuf. Technol.* 29, 97–101.

Rajurkar, K.P., Levy, G., Malshe, A., Sundaram, M.M., McGeough, J., Hu, X., Resnick, R., DeSilva, A., 2006. Micro and nano machining by electro-physical and chemical processes. *CIRP Ann. Manuf. Technol.* 55, 643–666. doi: 10.1016/j.cirp.2006.10.002

Rajurkar, K.P., Royo, G.F., 1989. Effect of RF control and orbital motion on surface integrity of EDM components. *J. Mech. Work. Technol.* 20, 341–352.

Rajurkar, K.P., Wang, W.M., Lindsay, R.P., 1989. A new model reference adaptive control of EDM. *CIRP Ann. Manuf. Technol.* 38, 183–186. doi: 10.1016/S0007-8506(07)62680-8.

Rajurkar, K.P., Wang, W.M., Lindsay, R.P., 1990. Real-time stochastic model and control of EDM. *CIRP Ann. Technol.* 39, 187–190.

Rasheed, M.S., Abidi, M.H., 2012. Analysis of influence of micro-EDM parameters on MRR, TWR and Ra in machining Ni-Ti shape memory alloy. *Int. J. Recent Technol. Eng.* 1, 32–37.

Sadeghzadeh, A., Asua, E., Feuchtwanger, J., Etxebarria, V., García-Arribas, A., 2012. Ferromagnetic shape memory alloy actuator enabled for nanometric position control using hysteresis compensation. *Sens. Actuators A Phys.* 182, 122–129.

Saxena, K.K., Agarwal, S., Khare, S.K., 2016. Surface characterization, material removal mechanism and material migration study of micro EDM process on conductive SiC. *Procedia CIRP* 42, 179–184. doi: 10.1016/j.procir.2016.02.267

Schubert, A., Zeidler, H., Hahn, M., Hackert-Oschätzchen, M., Schneider, J., 2013. Micro-EDM milling of electrically nonconducting zirconia ceramics. *Procedia CIRP* 6, 297–302. doi: 10.1016/j.procir.2013.03.026.

Schubert, A., Zeidler, H., Wolf, N., Hackert, M., 2011. Micro electro discharge machining of electrically nonconductive ceramics. *AIP Conf. Proc.* 1353, 1303–1308. doi: 10.1063/1.3589696.

Shabgard, M.R., Sadizadeh, B., Kakoulvand, H., 2009. The effect of ultrasonic vibration of workpiece in electrical discharge machining of AISIH13 tool steel. *Int. J. Mech. Aerosp. Ind. Mechatron. Manuf. Eng.* 3, 392–396.

Shah, A., Prajapati, V., Patel, P., Pandey, A., 2007. Development of pulsed power DC supply for micro-EDM. *UGC National Conference on Advances in Computer Integrated Manufacturing (NCACIM),* Department of Production and Industrial Engineering, J.N.V. University, Jodhpur.

Sheu, D.Y., Cheng, C.C., 2012. A hybrid microspherical styli gluing and assembling process on micro-EDM. *Mater. Manuf. Process.* 27, 1129–1134. doi: 10.1080/10426914.2012.677897.

Singh, P., Yadava, V., Narayan, A., 2018. Parametric study of ultrasonic-assisted hole sinking micro-EDM of titanium alloy. *Int. J. Adv. Manuf. Technol.* 94, 2551–2562. doi: 10.1007/s00170-017-1051-1.

Singh, R., 2010. Characterization of micro-EDM process for pyrolytic carbon 2010–2013. *Proceeding of the 7th International Conference on Micro Manufacturing,* Evanston, doi: 10.13140/2.1.1069.4089.

Snoeys, R., Cornelissen, H., 1975. Correlation between electro discharge machining data and machining settings. *Ann. CIRP* 24, 83–88.

Snoeys, R., Dauw, D.F., Kruth, J.P., Leuven, K.U., 1983. Survey of adaptive control in electro discharge machining. *J. Manuf. Syst.* 2(2), 147–164.

Song, K.Y., Chung, D.K., Park, M.S., Chu, C.N., 2010. Micro electrical discharge milling of WC-Co using a deionized water spray and a bipolar pulse. *J. Micromech. Microeng.* 20. doi: 10.1088/0960-1317/20/4/045022.

Song, X., Meeusen, W., Reynaerts, D., Van Brussel, H., 2000. Experimental study of micro-EDM machining performances on silicon wafer. *Micromach. Microfabr.* 331–339. doi: 10.1117/12.396450.

Suganthi, X.H., Natarajan, U., Sathiyamurthy, S., Chidambaram, K., 2013. Prediction of quality responses in micro-EDM process using an adaptive neuro-fuzzy inference system (ANFIS) model. *Int. J. Adv. Manuf. Technol.* 68, 339–347. doi: 10.1007/s00170-013-4731-5.

Tai, T.Y., Masusawa, T., Lee, H.T., 2007. Drilling microholes in hot tool steel by using micro-electro discharge machining. *Mater. Trans.* 48, 205–210. doi: 10.2320/matertrans.48.205.

Tak, H.S., Ha, C.S., Lee, H.J., Lee, H.W., Jeong, Y.K., Kang, M.C., 2011. Characteristic evaluation of Al2O3/CNTs hybrid materials for micro-electrical discharge machining. *Trans. Nonferrous Met. Soc. China,* English Ed. 21, S28–S32. doi: 10.1016/S1003-6326(11)61055-8.

Tamang, S.K., Natarajan, N., Chandrasekaran, M., 2017. Optimization of EDM process in machining micro holes for improvement of hole quality. *J. Braz. Soc. Mech. Sci. Eng.* 39, 1277–1287. doi: 10.1007/s40430-016-0630-7.

Tan, P.C., Yeo, S.H., 2011. Investigation of recast layers generated by a powder-mixed dielectric micro electrical discharge machining processg. *Proc. Inst. Mech. Eng. Part B J. Eng. Manuf.* 225, 1051–1062. doi: 10.1177/2041297510393645.

Tanabe, R., Ito, Y., Mohri, N., Masuzawa, T., 2016. Development of peeling tools with sub-50 μm cores by zinc electroplating and their application to micro-EDM. *CIRP Ann. Manuf. Technol.* 65, 221–224. doi: 10.1016/j.cirp.2016.04.025.

Tanabe, R., Ito, Y., Mohri, N., Masuzawa, T., 2011. Development of peeling tool for micro-EDM. *CIRP Ann. Manuf. Technol.* 60, 227–230. doi: 10.1016/j.cirp.2011.03.108.

Tarng, Y.S., Jang, J.L., 1996. Genetic synthesis of a fuzzy pulse discriminator in electrical discharge machining. *J. Intell. Manuf.* 7, 311–318. doi: 10.1007/BF00124831.

Tarng, Y.S., Tseng, C.M., Chung, L.K., 1997. A fuzzy pulse discriminating system for electrical discharge machining. *Int. J. Mach. Tools Manuf.* 37, 511–522. doi: 10.1016/S0890-6955(96)00033-8.

Teicher, U., Müller, S., Münzner, J., Nestler, A., 2013. Micro-EDM of carbon fibre-reinforced plastics. *Procedia CIRP* 6, 320–325. doi: 10.1016/j.procir.2013.03.092.

Ter Pey Tee, K., Hosseinnezhad, R., Brandt, M., Mo, J., 2013. Pulse discrimination for electrical discharge machining with rotating electrode. *Mach. Sci. Technol.* 17, 292–311. doi: 10.1080/10910344.2013.780559.

Thomas, N., Poongodi, P., 2009. Position control of DC motor using genetic algorithm based PID controller. *Proc. World Congr. Eng.* II, 1–5.

Tiwary, A.P., Pradhan, B.B., Bhattacharyya, B., 2018. Investigation on the effect of dielectrics during micro-electro-discharge machining of Ti-6Al-4V. *Int. J. Adv. Manuf. Technol.* 95, 861–874. doi: 10.1007/s00170-017-1231-z.

Tomura, M., Kunieda, M., 2010. Study on uncertainty of discharge energy in micro EDM. *Proceedings of the 4th CIRP International Conference on High Performance Cutting*, Gifu, Japan, 331–336.

Tong, H., Li, Y., Wang, Y., Yu, D., 2008. Servo scanning 3D micro-EDM based on macro/micro-dual-feed spindle. *Int. J. Mach. Tools Manuf.* 48, 858–869. doi: 10.1016/j.ijmachtools.2007.11.008.

Tong, H., Li, Y., Zhang, L., Li, B., 2013. Mechanism design and process control of micro EDM for drilling spray holes of diesel injector nozzles. *Precis. Eng.* 37, 213–221. doi: 10.1016/j.precisioneng.2012.09.004.

Trych-Wildner, A., Wildner, K., 2017. Multifilament carbon fibre tool electrodes in micro EDM—evaluation of process performance based on influence of input parameters. *Int. J. Adv. Manuf. Technol.* 91, 3737–3747. doi: 10.1007/s00170-017-0041-7.

Ubaid, A.M., Dweiri, F.T., Aghdeab, S.H., Al-Juboori, L.A., 2018. Optimization of electro discharge machining process parameters with fuzzy logic for stainless steel 304 (ASTM A240). *J. Manuf. Sci. Eng.* 140, 011013. doi: 10.1115/1.4038139.

Uhlmann, E., Bergmann, A., Bolz, R., Gridin, W., 2018. Application of additive manufactured tungsten carbide tool electrodes in EDM. *Procedia CIRP* 68, 86–90. doi: 10.1016/j.procir.2017.12.027.

Uhlmann, E., Roehner, M., 2008. Investigations on reduction of tool electrode wear in micro-EDM using novel electrode materials. *CIRP J. Manuf. Sci. Technol.* 1, 92–96. doi: 10.1016/j.cirpj.2008.09.011.

Uhlmann, E., Rosiwal, S., Bayerlein, K., Röhner, M., 2010. Influence of grain size on the wear behavior of CVD diamond coatings in micro-EDM. *Int. J. Adv. Manuf. Technol.* 47, 919–922. doi: 10.1007/s00170-009-2131-7.

Unune, D.R., Marani Barzani, M., Mohite, S.S., Mali, H.S., 2018. Fuzzy logic-based model for predicting material removal rate and average surface roughness of machined Nimonic 80A using abrasive-mixed electro-discharge diamond surface grinding. *Neural Comput. Appl.* 29, 647–662. doi: 10.1007/s00521-016-2581-4.

Vaezi, N., Tavakoli, P., Mosahehi, M., Borandeh, S.S., 2016. Improvement the performance of electrical discharge machining process with model reference adaptive controller. *Second International Congress on Technology, Communication and Knowledge, ICTCK 2015*, Mashhad, Iran, 570–575. doi: 10.1109/ICTCK.2015.7582731.

Venugopal, T.R., Rao, R., Prabhu, P., 2016. Piezoactuator based feed control system for in-situ tool grinding in micro-EDM. *J. Mech. Eng. Autom.* 6, 86–92. doi: 10.5923/c.jmea.201601.16.

Wan, Y., Kim, D.D., Park, Y., Joo, S., 2008. Micro electro discharge machining of polymethylmethacrylate (PMMA)/multi-walled carbon nanotube (MWCNT) nanocomposites. *Adv. Compos. Lett.* 17, 115–123.

Wang, A.C., Yan, B.H., Tang, Y.X., Huang, F.Y., 2005. The feasibility study on a fabricated micro slit die using micro EDM. *Int. J. Adv. Manuf. Technol.* 25, 10–16.

Wang, D., Zhao, W.S., Gu, L., Kang, X.M., 2011. A study on micro-hole machining of polycrystalline diamond by micro-electrical discharge machining. *J. Mater. Process. Technol.* 211, 3–11. doi: 10.1016/j.jmatprotec.2010.07.034.

Wang, J., Qian, J., Ferraris, E., Reynaerts, D., 2017. In-situ process monitoring and adaptive control for precision micro-EDM cavity milling. *Precis. Eng.* 47, 261–275. doi: 10.1016/j.precisioneng.2016.09.001.

Wang, J., Wang, Y.G., Zhao, F.L., 2009. Simulation of debris movement in micro electrical discharge machining of deep holes. *Mater. Sci. Forum* 626–627, 267–272. doi: 10.4028/www.scientific.net/MSF.626-627.267.

Wang, J., Yang, F., Qian, J., Reynaerts, D., 2016. Study of alternating current flow in micro-EDM through real-time pulse counting. *J. Mater. Process. Technol.* 231, 179–188. doi: 10.1016/j.jmatprotec.2015.12.010.

Wang, K., Zhang, Q., Liu, Q., Zhu, G., Zhang, J., 2017. Experimental study on micro electrical discharge machining of porous stainless steel. *Int. J. Adv. Manuf. Technol.* 90, 2589–2595. doi: 10.1007/s00170-016-9611-3.

Wang, W.M., 1988. A new EDM adaptive control plan using self-tuning control algorithm. *Proc. ASME Manuf. Int.* 1, 227–233.

Wang, W.M., Rajurkar, K.P., Akamatsu, K., 1995. Digital gap monitor and adaptive integral control for auto-jumping in EDM. *J. Eng. Ind.* 117, 253–258.

Wang, W.Y., Cheng, C.Y., Leu, Y.G., 2004. An online GA-based output-feedback direct adaptive Fuzzy-neural controller for uncertain nonlinear systems. *IEEE Trans. Syst. Man Cybern. Part B Cybern.* 34, 334–345. doi: 10.1109/TSMCB.2003.816995.

Weck, M., Dehmer, J.M., 1992. Analysis and adaptive control of EDM sinking process using the ignition delay time and fall time as parameter. *CIRP Ann. Technol.* 41, 243–246.

Xi, X.C., Chen, M., Zhao, W.S., 2017. Improving electrical discharging machining efficiency by using a Kalman filter for estimating gap voltages. *Precis. Eng.* 47, 182–190. doi: 10.1016/j.precisioneng.2016.08.003.

Xu, B., Wu, X., Lei, J., Cheng, R., Ruan, S., Wang, Z., 2015. Laminated fabrication of 3D queue micro-electrode and its application in micro-EDM. *Int. J. Adv. Manuf. Technol.* 80, 1701–1711. doi: 10.1007/s00170-015-7148-5.

Xu, B., Wu, X., Lei, J., Zhao, H., Liang, X., Cheng, R., Guo, D., 2018. Elimination of 3D micro-electrode's step effect and applying it in micro-EDM. *Int. J. Adv. Manuf. Technol.* 96, 429–438. doi: 10.1007/s00170-018-1632-7.

Yan, B.H., Wang, A.C., Huang, C.Y., Huang, F.Y., 2002. Study of precision micro-holes in borosilicate glass using micro EDM combined with micro ultrasonic vibration machining. *Int. J. Mach. Tools Manuf.* 42, 1105–1112. doi: 10.1016/S0890-6955(02)00061-5.

Yan, C., Zou, R., Yu, Z., Li, J., Tsai, Y., 2018. Improving machining efficiency methods of micro EDM in cold plasma jet. *Procedia CIRP* 68, 547–552. doi: 10.1016/j.procir.2017.12.111.

Yan, J., Tan, T.H., 2015. Sintered diamond as a hybrid EDM and grinding tool for the micromachining of single-crystal SiC. *CIRP Ann. Manuf. Technol.* 64, 221–224. doi: 10.1016/j.cirp.2015.04.069.

Yan, M.T., 2010. An adaptive control system with self-organizing fuzzy sliding mode control strategy for micro wire-EDM machines. *Int. J. Adv. Manuf. Technol.* 50, 315–328. doi: 10.1007/s00170-009-2481-1.

Yan, M.T., Chien, H.T., 2007. Monitoring and control of the micro wire-EDM process. *Int. J. Mach. Tools Manuf.* 47, 148–157. doi: 10.1016/j.ijmachtools.2006.02.006.

Yan, M.T., Fang, C.C., 2008. Application of genetic algorithm-based fuzzy logic control in wire transport system of wire-EDM machine. *J. Mater. Process. Technol.* 205, 128–137. doi: 10.1016/j.jmatprotec.2007.11.091.

Yan, M.T., Huang, K.Y., Lo, C.Y., 2009. A study on electrode wear sensing and compensation in micro-EDM using machine vision system. *Int. J. Adv. Manuf. Technol.* 42, 1065–1073. doi: 10.1007/s00170-008-1674-3.

Yan, M.T., Lin, S.S., 2011. Process planning and electrode wear compensation for 3D micro-EDM. *Int. J. Adv. Manuf. Technol.* 53, 209–219. doi: 10.1007/s00170-010-2827-8.

Yang, F., Qian, J., Wang, J., Reynaerts, D., 2018a. Simulation and experimental analysis of alternating-current phenomenon in micro-EDM with a RC-type generator. *J. Mater. Process. Technol.* 255, 865–875. doi: 10.1016/j.jmatprotec.2018.01.031.

Yang, J., Yang, F., Hua, H., Cao, Y., Li, C., Fang, B., 2018b. A bipolar pulse power generator for micro-EDM. *Procedia CIRP* 68, 620–624. doi: 10.1016/j.procir.2017.12.125.

Yang, G.Z., Liu, F., Lin, H.B., 2011. Research on an embedded servo control system of micro-EDM. *Appl. Mech. Mater.* 120, 573–577. doi: 10.4028/www.scientific.net/AMM.120.573.

Yang, Y., Mei, Y., He, T., 2010. Research of the micro-EDM discharge state detection method based on matlab fuzzy control. *2010 International Conference on Mechanic Automation and Control Engineering*, Wuhan, China, 3168–3171. doi: 10.1109/MACE.2010.5535610.

Yao, W.-S., Yang, F.-Y., Tsai, M.-C., 2011. Modeling and control of twin parallel-axis linear servo mechanisms for high-speed machine tools. *Int. J. Autom. Smart Technol.* 1, 77–85. doi: 10.5875/ausmt.v1i1.72.

Yaou, Z., Ning, H., Xiaoming, K., Wansheng, Z., Kaixian, X., 2017. Experimental study of an electrostatic field-induced electrolyte jet electrical discharge machining process. *Proc. Inst. Mech. Eng. Part B J. Eng. Manuf.* 231, 1752–1759. doi: 10.1177/0954405415612327.

Yeo, S.H., Aligiri, E., Tan, P.C., Zarepour, H., 2009a. A new pulse discriminating system for micro-EDM. *Mater. Manuf. Process.* 24, 1297–1305. doi: 10.1080/10426910903130164.

Yeo, S.H., Aligiri, E., Tan, P.C., Zarepour, H., 2009b. An adaptive speed control system for micro electro discharge machining. *AIP Conf. Proc.* 1181, 61–72. doi: 10.1063/1.3273682.

Yeo, S.H., Tan, P.C., Aligiri, E., Tor, S.B., Loh, N.H., 2009c. Processing of zirconium-based bulk metallic glass (BMG) using micro electrical discharge machining (Micro-EDM). *Mater. Manuf. Processes.* 24(12), 1242–1248.

Yeo, S.H., Kurnia, W., Tan, P.C., 2007a. Electro-thermal modelling of anode and cathode in micro-EDM. *J. Phys. D. Appl. Phys.* 40, 2513–2521. doi: 10.1088/0022-3727/40/8/015.

Yeo, S.H., Murali, M., 2003. A new technique using foil electrodes for the electro-discharge machining of micro grooves. *J. Micromech. Microeng.* 13. doi: 10.1088/0960-1317/13/1/401.

Yeo, S.H., Tan, P.C., Kurnia, W., 2007b. Effects of powder additives suspended in dielectric on crater characteristics for micro electrical discharge machining. *J. Micromech. Microeng.* 17. doi: 10.1088/0960-1317/17/11/N01.

Yoo, H.K., Ko, J.H., Lim, K.Y., Kwon, W.T., Kim, Y.W., 2015. Micro-electrical discharge machining characteristics of newly developed conductive SiC ceramic. *Ceram. Int.* 41, 3490–3496. doi: 10.1016/j.ceramint.2014.10.175.

Yoo, H.K., Kwon, W.T., Kang, S., 2014. Development of a new electrode for micro-electrical discharge machining (EDM) using Ti(C,N)-based cermet. *Int. J. Precis. Eng. Manuf.* 15, 609–616. doi: 10.1007/s12541-014-0378-x.

Yu, S.F., Lee, B.Y., Lin, W.S., 2001. Waveform monitoring of electric discharge machining by wavelet transform. *Int. J. Adv. Manuf. Technol.* 17, 339–343. doi: 10.1007/s001700170168.

Yu, Z.Y., Kozak, J., Rajurkar, K.P., 2003. Modelling and simulation of micro EDM process. *CIRP Ann. Manuf. Technol.* 52, 143–146. doi: 10.1016/S0007-8506(07)60551-4.

Yuangang, W., Fuling, Z., Jin, W., 2009. Wear-resist electrodes for micro-EDM. *Chinese J. Aeronaut.* 22, 339–342. doi: 10.1016/S1000-9361(08)60108-9.

Zhang, H., Zhang, J., Liu, H., Liang, T., Zhao, W., 2015. Dynamic modeling and analysis of the high-speed ball screw feed system. *Proc. Inst. Mech. Eng. Part B J. Eng. Manuf.* 229, 870–877. doi: 10.1177/0954405414534641.

Zhang, L., Jia, Z., Liu, W., Li, A., 2012. A two-stage servo feed controller of micro-EDM based on interval type-2 fuzzy logic. *Int. J. Adv. Manuf. Technol.* 59, 633–645. doi: 10.1007/s00170-011-3535-8.

Zhang, X., Shinshi, T., Endo, H., Shimokohbe, A., Imai, Y., Miyake, H., Nakagawa, T., 2007. Development of a 5-DOF controlled, wide-bandwidth, high-precision maglev actuator for micro electrical discharge machining. *Proceedings of the 2007 IEEE International Conference on Mechatronics and Automation, ICMA,* Harbin, China, 2877–2882. doi: 10.1109/ICMA.2007.4304016.

Zhang, X., Shinshi, T., Shimokohbe, A., 2008. High-speed electrical discharge machining by using a 5-DOF controlled maglev local actuator. *J. Adv. Mech. Des. Syst. Manuf.* 2, 493–503. doi: 10.1299/jamdsm.2.493.

Zhang, Y., 2005. The study on a new-type self-adaptive fuzzy logic control system in EDM process. *Proc. Fourth Int. Conf. Mach. Learn. Cybern.* 2, 726–730.

Zhang, Y., Chen, W., Cheng, H., Zhang, Y., 2018. Machinability for C/SiC composite material by electrical discharge machining. *Mater. Sci. Forum.* 913, 536–541. doi: 10.4028/www.scientific.net/MSF.913.536.

Zhang, Y., Zhang, W.J., Hesselbach, J., Kerle, H., 2006. Development of a two-degree-of-freedom piezoelectric rotary-linear actuator with high driving force and unlimited linear movement. *Rev. Sci. Instrum.* 77, 35112.

Zhou, M., Han, F., 2009. Adaptive control for EDM process with a self-tuning regulator. *Int. J. Mach. Tools Manuf.* 49, 462–469. doi: 10.1016/j.ijmachtools.2009.01.004.

Zhou, M., Han, F., Soichiro, I., 2008. A time-varied predictive model for EDM process. *Int. J. Mach. Tools Manuf.* 48, 1668–1677. doi: 10.1016/j.ijmachtools.2008.07.003.

Zhou, M., Wu, J., Xu, X., Mu, X., Dou, Y., 2018. Significant improvements of electrical discharge machining performance by step-by-step updated adaptive control laws. *Mech. Syst. Signal Process.* 101, 480–497. doi: 10.1016/j.ymssp.2017.06.041.

3

Micro EDM Milling

3.1 Introduction

One of the ways to fabricate 3D microfeatures using electro discharge machining (EDM) is to employ a tool with a complex geometry for machining. To fabricate those tools with intricate geometry, advanced tool processing methods have to be used. This increases the tool fabrication time and total cost of machining. Frequent tool changes are also required, as the microfeature on the tool may wear out due to tool erosion. Considering this, using a simple-shaped tool to fabricate 3D micromolds and other complex microstructures by controlling the tool path can be considered as an efficient alternative (Rajurkar and Yu, 2000). Micro EDM milling is a variant of EDM with simple tooling, more flexibility, and comparatively complex machining strategy. In micro EDM milling, the tool electrode rotates continuously and follows a particular tool path to carve out the designed shape. Commercially available cylindrical rods or on-machine fabricated electrodes of simple shape can be used as the tool electrode. Readily available microrods reduce the tool processing time substantially.

Contrary to EDM drilling and die-sinking EDM, the tool moves in all directions (vertical and horizontal plane), and the effect of wear at the bottom area as well as the peripheral area will be predominant. Changing the tools during machining to compensate the wear-related errors is impractical as the tool positioning system may fail to place the tool at the same location every time after a tool change. To ensure precision machining, appropriate tool wear compensation strategies or machining strategies need to be implemented. Rotating tool electrode in EDM milling will create a self-centering effect on the tool electrodes, which reduces the effect of tool deflections on the microfeature shape (Richard and Demellayer, 2013).

3.2 Principle of Machining

Two types of machining strategies used in micro EDM milling are (1) bulk machining and (2) layerwise machining.

3.2.1 Bulk Machining

In this, the machining process is completed with a single pass. For compensating the tool wear, the trend of tool wear is analyzed experimentally, analytically, and using simulation tools. Later, a "corrected tool path" is generated to follow during milling operation (Meeusen et al., 2002; Pei et al., 2016). Figure 3.1a shows the method of bulk machining with a modified tool path for wear compensation. As the effect of corner wear persists in this method, it will introduce a tapering effect in the final microfeature.

3.2.2 Layerwise Machining

The microfeature geometry is sliced into layers, and the machining is carried out in a layer-by-layer manner. Usually, the principle of uniform wear method (UWM) is used to determine the thickness of the layer and the tool path direction. The tool follows the designed tool path and periodically moved down at the end of each "layer machining cycle." By this method, the effect of corner wear disappears as the tool retains the flatness at the end of each layer. A computer aided design and manufacturing (CAD/CAM) tool is an essential part of this strategy to generate complex tool paths (Li et al., 2007). An advanced servo control mechanism can be used to maintain optimum gap width and optimum layer thickness. With the help of advanced CAM systems as well as sophisticated

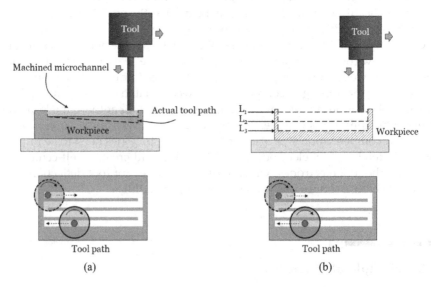

FIGURE 3.1
Machining strategy in micro EDM milling: (a) layerwise machining and (b) bulk machining.

servo control and positioning mechanism, scanning micro EDM is evolved with the capability of machining very complex 3D shapes with high surface quality (Tong et al., 2008, 2016, 2018). Figure 3.1b shows a schematic representation of layerwise machining strategy, where L_1, L_2, and L_3 are the machining layers.

3.3 Microscopic Analysis of the Machining Process

Being a variant of micro EDM, the material removal mechanism is considered to be melting and vaporization due to electrical discharges. The machining zone can be divided into two regions as shown in Figure 3.2. Unlike EDM drilling, the peripheral discharges remove a significant portion of the workpiece material in EDM milling. The rotation of the EDM tool electrode with continuous translational movement along the workpiece plane introduces some new trends in machining behavior. Karthikeyan et al. (2012) studied the machining behavior of micro EDM milling at a microscopic scale. The study focused on the effects of tool rotation and discharged energy on the surface characteristics of microchannels. The main difference in machining behavior is considered to be the smashing of globules due to different forces, change in nature of the resolidified layer, and formation of milling marks on the channel bottom surface.

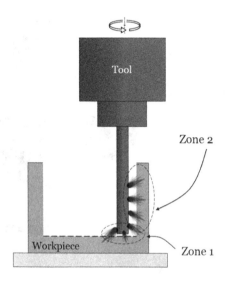

FIGURE 3.2
Different machining zones in micro EDM milling.

3.3.1 Effect of Tool Rotation on Sparking

The microtool electrodes may experience periodic deflections due to the shock waves generated from discharge plasma collapsing. These deflections will lead to unexpected short circuits and "welding" of tool electrode to the workpiece surface. Tool rotation helps to impart more efficient dielectric flushing, tool stability, and self-centering of the tool electrode. Also, tool rotation will generate more uniform sparks and overlapping craters on the workpiece surface, which improves the machined surface quality.

3.3.2 Globule Formation

The molten material that is thrown out of the machined crater after the pulse-ON period cools down due to interaction with fresh dielectric fluid in the vicinity. This quenching effect will result in the formation of non-crystalline globules. The size and number of these spherical particles may depend on the discharge energy and rotation speed. Figure 3.3 shows the globules formed during micro EDM milling with different discharge

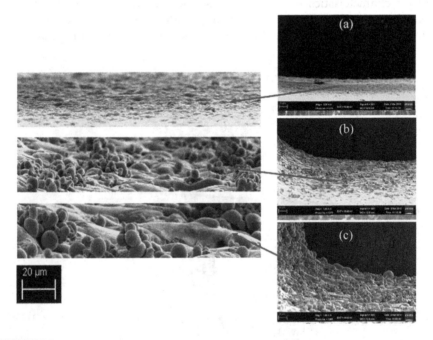

FIGURE 3.3
Globule formation during micro EDM milling for different discharge energy levels: (a) 50 μJ, (b) 500 μJ, and (c) 2,000 μJ (Karthikeyan et al., 2012).

energy levels. From Figure 3.3, nonuniform size distribution among the globules is observed. The uneven size distribution may be due to merging or separation of particles while moving in the rotating dielectric fluid. However, the average size of the globules increases with discharge energy. The particles experience two kinds of forces during micro EDM milling: (1) frictional force between the stirring dielectric fluid and channel walls and (2) centrifugal force that throws the particles away from the tool surface towards the side walls (Karthikeyan et al., 2012). The action of these forces results in smooth depositions of the globules in a semisolid state on the channel surface.

3.3.3 Formation of Milling Marks

Figure 3.4a shows the circular patterns at the entrance section of the tool indicating the flow of molten metal during the machining process. Figure 3.4b shows the milling marks on the bottom surface of the microchannel. Unlike the conventional end milling operation, the milling marks are formed due to the stirring action of molten material in the vicinity of the rotating tool electrode. The stirring molten material in the tool bottom region interacts with the channel surface continuously, resulting in circular marks. However, these marks are partially removed due to subsequent sparks, which transform into broken arc-shaped marks on the machined surface. As the discharge energy increases, the amount of material thrown out from the craters increases and the milling marks are replaced by a thin resolidified layer (Karthikeyan et al., 2012).

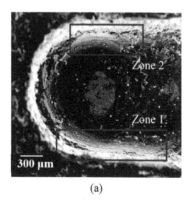
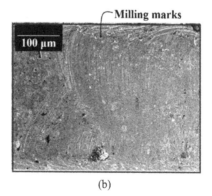

(a)   (b)

FIGURE 3.4
(a) Effect of rotation of tool electrode on molten metal flow during microchannel machining using EDM and (b) milling marks on the microchannel bottom during machining with low discharge energy (Karthikeyan et al., 2012).

3.4 Tool Path Generation

As the complexity of the microfeature increases, the need for a CAD/CAM system for tool path modeling becomes inevitable. Moreover, in micro EDM milling, prediction and correction of the errors due to tool wear must be incorporated into the CAD/CAM tool. Early work on developing a CAD/CAM environment for micro EDM milling by incorporating tool wear compensation strategies (based on UWM) is done by Rajurkar and Yu (2000). The applicability of the system is proven by machining a cavity with a pyramidal micro features at the center. Meeusen et al. (2002) developed a CAD/CAM system based on a prediction of the machined profile curve by generating a complimentary tool path for correcting the errors.

Because of the mode of machining and tool electrode shape, the existence of some vestige on the machined surface was evident, particularly for complex-shaped features where the machining profile cannot be defined easily with normal equations. Zhao et al. (2004) developed a slicing strategy for micro EDM milling in which isoparametric interpolation is employed for generating a tool path for free-form surfaces. The general method of slicing strategy for micro EDM milling is shown in Figure 3.5. CAD/CAM system for micro EDM milling is improved by utilizing contour-parallel scanning path (Tong et al., 2007). Machining curves by short line interpolation and point-by-point comparison are developed as part of the CAD/CAM tool. To realize a reliable simulation system for predicting machining profile, a micro EDM milling virtual simulator is developed by Heo et al. (2009) using a Z-map algorithm. A new interpolation method with variable period and step size based on

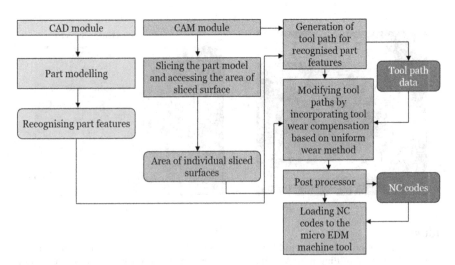

FIGURE 3.5
Strategy for generating tool paths for layer-by-layer machining in micro EDM milling.

square constraint is put forward by Wei et al. (2012), which improved the machining efficiency by 30% and reduced the volume of numerical control (NC) codes by 12.5%.

3.5 Process Capability and Applications

The ability of micro EDM milling to machine complex microstructures in difficult to machine materials makes it a prominent precision machining process for fabrication of micromolds and other 3D microstructures. Some of the main applications of micro EDM milling are as follows.

3.5.1 Fabrication of 3D Microfeatures

- Silicon microstructures for microsensors are fabricated by micro EDM milling with an aspect ratio up to 20 (Song et al., 1999).
- Using an advanced CAD/CAM tool, a human face is carved in a cavity of 1 mm diameter (Figure 3.6a) by Zhao et al. (2004).
- High-aspect-ratio rectangular protrusions are fabricated on the aluminum surface (Lim et al., 2003). Reverse polarity is reported to be suitable for aluminum machining to avoid the formation of a nonconductive oxide layer.
- Based on a tool wear compensation strategy considering the latent heat of fusion and evaporation, a microplatinum hemisphere is machined with high surface quality (Hang et al., 2006).
- Rectangular microfeatures are machined to study the effect of resistivity of deionized water as a dielectric fluid (Chung et al., 2007).
- Micromold inserts for compression molding is fabricated using EDM milling (Cao et al., 2007). The inserts are tested by fabricating high-aspect-ratio microcomponents of Al (Figure 3.6b) by microcasting.
- Asad et al. (2007) and Tong et al. (2007) fabricated various microstructures including micropyramids and microchannels on the metal surface (Figure 3.6c).
- Complex microstructure and cavities of various shapes (eye-shaped microstructure (Figure 3.6d), protruding micro hexaprism (Figure 3.6e), micro hexaprism cavity, micro convex-hemisphere, and multi microtriangles) are machined by multilayer scanning EDM with the help of micro dual-feed spindle (Tong et al., 2008, 2016, 2018).

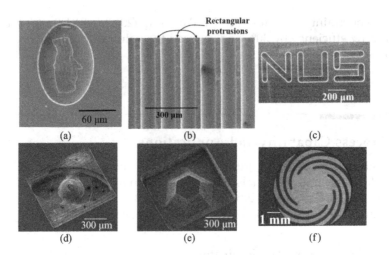

FIGURE 3.6
Micro EDM milling applications: (a) machining a human face with the help of special CAD/
CAM system (Zhao et al., 2004), (b) high-aspect-ratio rectangular protrusions on aluminum
(Cao et al., 2007), (c) carving letters with tool electrodes of size less than 20 μm (Asad et al.,
2007), (d) eye-shaped microstructure (Tong et al., 2008), (e) micro hexaprism (Tong et al., 2008),
and (f) thrust air-bearing surface with circular microgrooves (Ferraris et al., 2011).

- Ferraris et al. (2011) machined different microcomponents using micro EDM milling, including thrust air-bearing surface with circular grooves (Figure 3.6f)
- Micro EDM milling is used to machine 3D conical diffuser holes in turbine blades of civil engines to provide efficient cooling. A seven-axis micro EDM is used for machining of cooling paths with different dimensions. Case study: Aerospace application, www.sarix.com).

3.5.2 Microcomponents and Microfluidic Devices

- Micro EDM milling is proposed as a cheap alternative to the lithographic process to fabricate biocompatible microfluidic devices with high surface quality ($R_a = 0.4$ μm), to use for implant abutment (Murali and Yeo, 2004). Microchannels machined on TiAl6V using EDM for a biomedical device are shown in Figure 3.7a.
- Micro Swiss-roll combustor with burr-free channels is fabricated on beryllium–copper alloy (Ali, 2009).
- Micro EDM milling is utilized for fabricating turbine impeller for micro power generation system on Si_3N_4-TiN ceramic composite (Liu et al., 2010). The impeller is fabricated using layer-by-layer processing employing tungsten carbide tool electrodes of 1 mm diameter for roughing and 0.7 mm electrodes for finishing (Figure 3.7b).

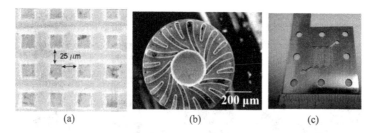

FIGURE 3.7
Micro EDM milling for fabrication of 3D microcomponents: (a) biocompatible microfluidic device fabricated on Ti-6Al-4V (Murali and Yeo, 2004), (b) microturbine impeller for power generation system (Liu et al., 2010), and (c) bipolar plates with microchannels used in fuel cells (Hung et al., 2011).

- In a fuel cell, bipolar plates are responsible for 60%–80% of the stack weight and 50% of the stack volume (Hung et al., 2011). Metallic bipolar plates for fuel cell applications are processed with micro EDM milling as shown in Figure 3.7c. It is reported that high-aspect-ratio, burr-free microchannels fabricated by micro EDM improve the fuel cell performance compared to the channels processed with electroforming or electrochemical machining (Hung et al., 2011).

- Micromixers have an essential role in microfluidics, enabling rapid mixing of reactants. A staggered herringbone micromixer is fabricated using micro EDM milling with a straight groove of 90 µm width and staggered herringbone groove of 70 µm width (Sabotin et al., 2019).

- Micronozzles have a crucial role in micropropulsion systems for accelerating the gas flow. A segmented milling technique using micro EDM is used to make a micronozzle on ZrB_2–SiC–graphite ceramic (Li et al., 2018). A multistep process employing tool electrodes of different diameters is used to machine the nozzle.

3.5.3 Micromolds and Punching Dies

Micro EDM is popularly used for fabrication of micromolds for injection molding.

- Yu (1997) machined a microcar mold of dimensions $0.5 \times 0.2 \times 0.2$ mm and micro half ball molds (www.mech.kuleuven.be/en/research/mpe) using milling EDM.

- A spherical micromold (Figure 3.8a) is fabricated using micro EDM milling by Meeusen et al. (2016). Two different tool path designs are used for mold fabrication. The mold is tested by fabricating

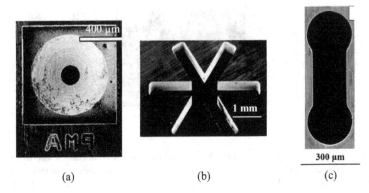

FIGURE 3.8
Fabrication of micromolds and dies using micro EDM milling: (a) spherical micromold for injection molding, (b) extrusion die for fiber optics application (Ferraris et al., 2011), and (c) microdie for punch fabrication (Yu et al., 2019).

poly-methyl methacrylate (PMMA) microstructures using hot embossing and injection molding arrangement.

- Ferraris et al. (2011) machined a microextrusion die (Figure 3.8b) for extrusion of fiber optics.

- A micromold for biomedical application with a continuous radius profile on a hemispherical surface with an inclined chipped-out face is fabricated by micro EDM milling. A dedicated CAM system ensured the dimensional accuracy of the slots and other features in the micromold. (Case study: medical application, www.sarix. com)

- Micromolds for producing microfluidic devices are fabricated using EDM. The protrusions in the micromold have a width of $10\,\mu m$ and a height of $100\,\mu m$. The dimensional accuracy is maintained within $1\,\mu m$, and the surface roughness is kept below $80\,nm$. (Case study: IMTEK, www.sarix.com)

- A microdie is fabricated by micro EDM milling and is used to make a micropunch by reverse EDM strategy (Yu et al., 2019; Zeng et al., 2018), as shown in Figure 3.8c.

3.6 Advantages

- The need for complex tooling is avoided.
- Commercially available rods of standard size can be used as the tool electrode.

- Simultaneous control of the axes using computer numerical control (CNC) systems helps to fabricate 3D complex microstructures with high precision.
- The effect of tool deflections can be avoided by rotating the tool electrode at high speed, which imparts a self-centering effect on the tool (Richard and Demellayer, 2013).
- Development of models and simulations of material removal and tool wear helps to generate efficient tool path designs for minimum dimensional deviations (Karthikeyan et al., 2011).
- With the help of CAD/CAM systems, tool paths for complex shape can be generated easily (Richard and Giandomenico, 2018; Heo et al., 2009; Zhang et al., 2015).
- Tool rotation facilitates dielectric flushing, debris removal, and uniformly distributes the discharges (Karthikeyan et al., 2012).
- Overlapping craters makes shallow elongated craters and increases surface quality (Karthikeyan et al., 2012).
- Can be used to make microchannels on nonconductive materials with the help of an additional conductive layer on the workpiece surface (Schubert et al., 2013).
- Compared to microend milling process, very thin microfeatures can be fabricated without any bending or buckling.

3.7 Challenges

- Predominant tool wear in all directions reduces the geometrical and dimensional accuracy.
- More complex servo control strategy compared to EDM drilling is required due to motion in all directions (Li et al., 2007).
- Tool path design is intricate due to the addition of tool wear compensation methods.
- Predominant recast layer formation on the surface.
- Unavailability of adequate models for material removal mechanism makes the prediction of tool wear and material removal rate less accurate.

Tristo et al. (2015) conducted an energy efficiency study of micro EDM milling and reported that only 3% of the total energy consumption is utilized for material removal. The study suggested that the appropriate use of auxiliary attachments would reduce the energy consumption and increase the energy efficiency of micro EDM milling.

3.8 Conclusion

Compared to other microfabrication methods for 3D microfeatures, micro EDM milling needs simple tooling and fixture. With the help of advanced CNC systems, any complex shapes can be processed using this method. However, the tool electrode erosion during machining demands advanced tool wear compensation strategies to be incorporated during machining. Compared to EDM drilling or die-sinking EDM, the servo control has to be more advanced to control harmful sparks from all directions. Micro EDM milling is widely used in structuring, texturing, and mold-making operations to make components of sophisticated features and burr-free surfaces.

References

Ali, M.Y., 2009. Fabrication of microfluidic channel using micro end milling and micro electrical discharge milling. *Int. J. Mech. Mater. Eng.* 4, 93–97.

Asad, A.B.M.A., Masaki, T., Rahman, M., Lim, H.S., Wong, Y.S., 2007. Tool-based micro-machining. *J. Mater. Process. Technol.* 192–193, 204–211.

Cao, D.M., Jiang, J., Meng, W.J., Jiang, J.C., Wang, W., 2007. Fabrication of high-aspect-ratio microscale Ta mold inserts with micro electrical discharge machining. *Microsyst. Technol.* 13, 503–510.

Chung, D.K., Kim, B.H., Chu, C.N., 2007. Micro electrical discharge milling using deionized water as a dielectric fluid. *J. Micromech. Microeng.* 17, 867–874.

Ferraris, E., Reynaerts, D., Lauwers, B., 2011. Micro-EDM process investigation and comparison performance of Al $3O_2$ and ZrO_2 based ceramic composites. *CIRP Ann. Manuf. Technol.* 60, 235–238.

Hang, G., Cao, G., Wang, Z., Tang, J., Wang, Z., Zhao, W., 2006. Micro-EDM milling of micro platinum hemisphere. *2006 1st IEEE International Conference on Nano/Micro Engineered and Molecular Systems, IEEE-NEMS*, Bangkok, Thailand, 579–584.

Heo, S., Jeong, Y.H., Min, B.K., Lee, S.J., 2009. Virtual EDM simulator: Three-dimensional geometric simulation of micro-EDM milling processes. *Int. J. Mach. Tools Manuf.* 49, 1029–1034.

Hung, J.C., Yang, T.C., Li, K.C., 2011. Studies on the fabrication of metallic bipolar plates - using micro electrical discharge machining milling. *J. Power Sources* 196, 2070–2074. doi:10.1016/j.jpowsour.2010.10.001.

Karthikeyan, G., Garg, A.K., Ramkumar, J., Dhamodaran, S., 2012. A microscopic investigation of machining behavior in μED-milling process. *J. Manuf. Process.* 14, 297–306.

Karthikeyan, G., Sambhav, K., Ramkumar, J., Dhamodaran, S., 2011. Simulation and experimental realization of μ-channels using a μeD-milling process. *Proc. Inst. Mech. Eng. Part B J. Eng. Manuf.* 225, 2206–2219.

Li, H., Wang, Y., Wang, Z., Zhao, Z., 2018. Fabrication of ZrB 2 – SiC – graphite ceramic micro-nozzle by micro-EDM segmented milling. *J. Micromech. Microeng.* 28(10), 105022.

Li, Y., Tong, H., Cui, J., Wang, Y., 2007. Servo scanning EDM for 3D micro structures. *First International Conference on Integration and Commercialization of Micro and Nanosystems Parts A B,* Hainan, China, 1369–1374.

Lim, H.S., Wong, Y.S., Rahman, M., Edwin Lee, M.K., 2003. A study on the machining of high-aspect ratio micro-structures using micro-EDM. *J. Mater. Process. Technol.* 140, 318–325.

Liu, K., Lauwers, B., Reynaerts, D., 2010. Process capabilities of micro-EDM and its applications. *Int. J. Adv. Manuf. Technol.* 47, 11–19.

Meeusen, W., Reynaerts, D., Peirs, J., van Brussel, H., Dierickx, V., Driesen, W., 2016. The machining of freeform micro moulds by micro EDM ; work in progress. *Proceedings of the 12th Micromechanics Europe Workshop MME 2001,* Cork, Ireland, 46–49.

Meeusen, W., Reynaerts, D., van Brussel, H., 2002. CAD tool for the design and manufacturing of freeform micro-EDM electrodes. *Symposium on Design, Test, Integration, and Packaging of MEMS/MOEMS 2002,* Mandelieu-La Napoule, France, 105–113.

Murali, M., Yeo, S.H., 2004. Rapid biocompatible micro device fabrication by micro electro-discharge machining. *Biomed. Microdevices* 6, 41–45.

Pei, J., Zhou, Z., Zhang, L., Zhuang, X., Wu, S., Zhu, Y., Qian, J., 2016. Research on the equivalent plane machining with fix-length compensation method in micro-EDM. *Procedia CIRP* 42, 644–649.

Rajurkar, K.P., Yu, Z.Y., 2000. 3D micro-EDM using CAD/CAM. *CIRP Ann. Manuf. Technol.* 49, 127–130.

Richard, J., Demellayer, R., 2013. Micro-EDM-milling development of new machining technology for micro-machining. *Procedia CIRP* 6, 292–296.

Richard, J., Giandomenico, N., 2018. Electrode profile prediction and wear compensation in EDM-milling and. *Procedia CIRP* 68, 819–824.

Sabotin, I., Tristo, G., Lebar, A., Jerman, M., Prijatelj, M., Drešar, P., Valentinčič, J., 2019. Preliminary study on staggered herringbone micromixer design suitable for micro EDM milling, in: S. Hloch, D. Klichová, G. Krolczyk, S. Chattopadhyaya, L. Ruppenthalová, eds. *Advances in Manufacturing Engineering and Materials.* Springer, Switzerland, pp. 229–236.

Schubert, A., Zeidler, H., Hahn, M., Hackert-Oschätzchen, M., Schneider, J., 2013. Micro-EDM milling of electrically nonconducting zirconia ceramics. *Procedia CIRP* 6, 297–302.

Song, X., Reynaerts, D., Meeusen, W., van Brussel, H., 1999. Micro-EDM for silicon microstructure fabrication. *Proceedings of Design, Test, and Microfabrication of MEMS and MOEMS.* International Society for Optics and Photonics, 3680, 792–800.

Tong, H., Cui, J., Li, Y., Wang, Y., 2007. CAD/CAM integration system of 3D micro EDM. *2007 First International Conference on Integration and Commercialization of Micro and Nanosystems,* Sanya, China. American Society of Mechanical Engineers, 1383–1387.

Tong, H., Li, Y., Wang, Y., Yu, D., 2008. Servo scanning 3D micro-EDM based on macro/micro-dual-feed spindle. *Int. J. Mach. Tools Manuf.* 48, 858–869.

Tong, H., Li, Y., Zhang, L., 2016. On-machine process of rough-and-finishing servo scanning EDM for 3D micro cavities. *Int. J. Adv. Manuf. Technol.* 82, 1007–1015.

Tong, H., Li, Y., Zhang, L., 2018. Servo scanning 3D micro EDM for array micro cavities using on-machine fabricated tool electrodes. *J. Micromech. Microeng.* 28, 1–8.

Tristo, G., Bissacco, G., Lebar, A., Valentinčič, J., 2015. Real time power consumption monitoring for energy efficiency analysis in micro EDM milling. *Int. J. Adv. Manuf. Technol.* 78, 1511–1521.

Wei, L., Zhang, L., Liu, W., Jia, Z., 2012. A new interpolation method of variable period and step size in micro-EDM milling based on square constraint. *Int. J. Adv. Manuf. Technol.* 63, 621–629.

Yu, Z., 1997. Three dimensional micro-EDM using simple electrodes. Disseration University of Tokyo, 149–155.

Yu, Z., Li, D., Yang, J., Zeng, Z., Yang, X., Li, J., 2019. Fabrication of micro punching mold for micro complex shape part by micro EDM. *Int. J. Adv. Manuf. Tech.* 100(1–4), 743–749.

Zeng, Z., Li, D., Yu, Z., Yang, X., Li, J., Kang, R., 2018. Study of machining accuracy of micro punching mold using micro-EDM. *Procedia CIRP* 68, 588–593.

Zhang, L., Du, J., Zhuang, X., Wang, Z., Pei, J., 2015. Geometric prediction of conic tool in micro-EDM milling with fix-length compensation using simulation. *Int. J. Mach. Tools Manuf.* 89, 86–94.

Zhao, W., Yang, Y., Wang, Z., Zhang, Y., 2004. A CAD/CAM system for micro-ED-milling of small 3D freeform cavity. *J. Mater. Process. Technol.* 149, 573–578.

4

Micro Die-Sinking EDM

4.1 Introduction

Among the variants of electro discharge machining (EDM) machines, die-sinking EDM is considered to be the oldest one that utilizes electrical discharges for controlled material removal. Die-sinking EDM machines demand less complex axis control because the automated motion is mostly restricted to a single axis. However, the tooling is more complex compared to any other variants of EDM. As the name suggests, the principle of die-sinking EDM is similar to shaping metal sheets using a die and a press. Instead of shaping the metal with mechanical force, die-sinking EDM utilizes electrical discharges to replicate the complementary image of the die on the workpiece surface as shown in Figure 4.1. The tool is not allowed to rotate but advanced to the workpiece surface, keeping a constant spark gap in between.

Compared to other variants of EDM, die-sinking EDM machines are popular for being less expensive. Simpler axis controllers, nonrotating spindles, and absence of other complex auxiliary systems make this EDM variant affordable to a medium-level manufacturing enterprise. To fabricate the EDM

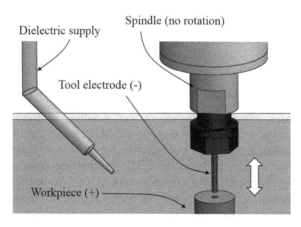

FIGURE 4.1
Schematic representation of micro die-sinking EDM.

tool electrode, die-sinking EDM utilizes different processes such as end milling, etching, forming, etc. Micro die-sinking EDM can be considered as the scaled-down version of conventional die-sinking EDM technology. Even though the machinable feature dimensions are scaled down to the microregime, the risks associated with the process are scaled up due to the increase of process unpredictability. The tool fabrication processes have to be ultraprecise to ensure the dimensional accuracy of the finished microcomponent. Usually, the die-pressing operation is considered to be quicker than the conventional milling for fabricating complex 3D shapes, excluding the time taken for the die preparation. Similarly, the die-sinking operation can be faster than the EDM milling operation to fabricate free-form surfaces as the tool only has to plunge to the workpiece instead of following complex tool paths.

4.2 Principle of Operation

As the tool with a specific geometry is plunged into the workpiece, the dielectric fluid fills the spark gap, and the first spark occurs after a dielectric breakdown at the minimum resistance point where the gap is minimum. Due to material removal from tool and workpiece, this first spark point will not be the minimum distance point after that, and the next spark ablates material from a different point. As the tool further advances towards the workpiece, more material is removed from the workpiece surface by exactly replicating the negative image of the tool shape. To enhance the dielectric flow, the tool electrode is allowed to move back and forth like a peck drilling operation. This action ensures the availability of fresh dielectric fluid in the spark gap and removes molten material from the machining zone.

4.3 Machining Behavior

4.3.1 Secondary Discharges

Very small spark gap and absence of tool rotation reduce the dielectric flushing efficiency in micro die-sinking EDM. A reciprocating motion of the tool is essential as the depth of machining increases. The reciprocating motion introduces some turbulence in the machining zone and enhances debris flushing. However, the effectiveness of tool jumping depends on the height of jumping. As the height of tool jumping increases, the debris removal is observed to be more efficient. However, the tool jumping time is a nonproductive time, and the total cycle time increases with jump height. The floating debris particles will be electrostatically polarized and initiate secondary discharges (Cao et al., 2010). These discharges affect the tool shape and microfeature dimensions.

4.3.2 Workpiece Deposition

During micro EDM using hydrocarbon oil as a dielectric fluid, the formation of a thin carbon layer on the tool surface is common. Some researchers reported this deposition as advantageous regarding reduction in effective tool electrode wear. In some cases, deposition of workpiece material is also found on the tool surface, and this phenomenon is reported to be predominant in the case of micro die-sinking EDM. Interestingly, these depositions are more concentrated in the nonworking zones or the area far from the tool bottom (primary machining zone). Murray et al. (2012) studied this phenomenon using a reciprocating micro EDM setup. W-Cu alloy is used as the tool electrode and Co-Cr alloy as the workpiece material. Figure 4.2 shows that the size of the deposition increases as it moves far from the primary discharge zone. This study revealed that the deposition on the tool electrode surface is not due to the subsequent solidification of the molten pool thrown out from the primary discharge zone onto the tool surface. The molten material from the discharge crater is washed away by the dielectric fluid which solidifies it to a granular form. These debris particles move along with the fluid and ideally, should be washed away from the machining region. However, in die-sinking micro EDM, the debris evacuation is very challenging due to the absence of tool rotation and a very small spark gap. The thrown-out particles from the primary machining zone get accumulated in the nonworking zones and initiate secondary sparks at various points. These secondary sparks will remelt the debris particles, and the melted material will be deposited on the tool electrode surface as shown in Figure 4.3. As this process continues, a layer of workpiece material with a thickness ranging from 1 to 3 μm is formed on the tool electrode surface as shown in Figure 4.4. Even though this

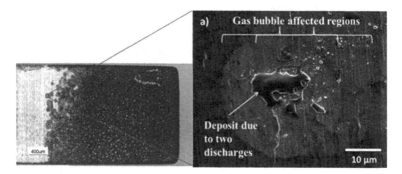

FIGURE 4.2
Microscopic analysis of workpiece deposition on tool surface during micro die-sinking EDM (Murray et al., 2012).

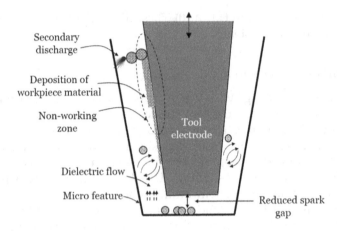

FIGURE 4.3
Deposition of workpiece material on the tool surface.

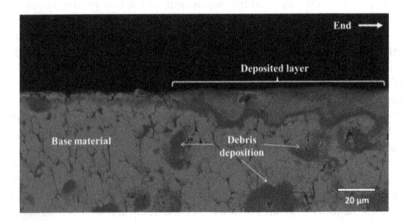

FIGURE 4.4
Workpiece deposition layer formed on the tool surface away from the primary machining zone (Murray et al., 2012).

layer changes the effective dimensions of the tool electrode, the tool wear erosion from the parent metal is reduced due to this covering. Due to the lack of rotation, the accumulation may occur at random points and will result in a nonuniform coating.

4.3.3 Step Effect

Three-dimensional die-sinking electrodes fabricated via superimposing 2D microstructures (Xu et al., 2015) show steps on the final electrode surface, which is known as the stepping effect. When those electrodes are used

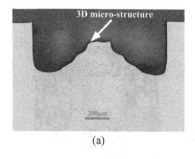
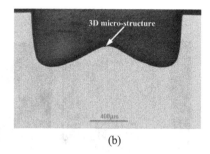

(a)   (b)

FIGURE 4.5
(a) Step effect during machining with 3D microelectrodes and (b) smoothened microfeature surface with the help of skin effect and point discharge (Xu et al., 2018b).

in machining, the steps in the electrode will be replicated on the microfeature surface as shown in Figure 4.5a. This will affect the surface quality and dimensional accuracy of the feature. Xu et al. (2018b) developed a strategy to eliminate these steps from microfeature surface using skin effect. When high-frequency pulses are used in micro EDM, the eddy current counteracts with the current flow and the electromagnetic field distribution changes. As the maximum field density shifts to the outer edges of the electrodes, the discharge probability will be maximum at tool edges (Schacht et al., 2004; Liu et al., 2017). The increased number of discharges at the outer edges leads to erosion of tool material and curving of tool edges. Xu et al. (2018b) make use of this phenomenon to fabricate smooth microfeatures using stepped 3D die-sinking EDM electrode. Due to skin effect, the steps in the tool electrode are worn out easily as the machining progresses, and a step-free microfeature surface is obtained as shown in Figure 4.5b.

4.4 Applications and Process Capabilities

4.4.1 Fuel Cell

Bipolar plates are an integral part of proton exchange membrane fuel cells. They consist of high-aspect-ratio 3D microchannels. Hung et al. (2012) used micro die-sinking EDM with a 3D microtool electrode (fabricated by high-speed milling) to make a bipolar plate as shown in Figure 4.6. Even though machining time (excluding the tool fabrication time) is less than the micro EDM milling operation, the machined surface quality is comparatively low (Hung et al., 2012).

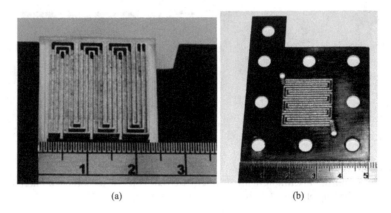

(a)   (b)

FIGURE 4.6
(a) Tool electrode for micro die-sinking EDM and (b) fabricated bipolar plate (Hung et al., 2012).

4.4.2 Micromixing Device

Micro die-sinking EDM can be used to fabricate micromixing devices with the help of 3D electrodes fabricated by ultraprecision mechanical milling operation as shown in Figure 4.7 (Uhlmann et al., 2005).

4.4.3 3D Microstructures

Micro die-sinking EDM is widely used for fabricating microstructures on electrically conductive workpiece materials. Figure 4.8 shows a die-sinking EDM electrode and a fabricated microstructure with complex geometry

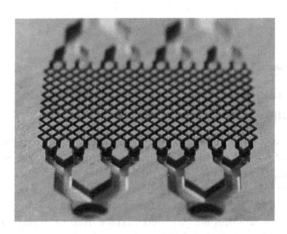

FIGURE 4.7
Micro die-sinking EDM electrode for micromixer (Uhlmann et al., 2005).

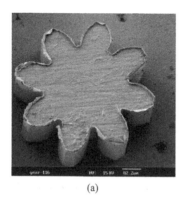
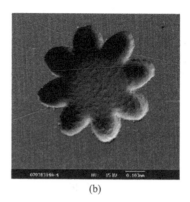

(a)   (b)

FIGURE 4.8
(a) EDM die-sinking electrode for microstructuring and (b) fabricated microstructure (Tong et al., 2008).

(Tong et al., 2008). Three-dimensional microstructures are also fabricated with foil-type electrodes (Xu et al., 2018a). Foil-type electrodes overcome the need for highly complex tool electrode fabrication methods. However, the step effect will be prevalent during machining.

4.4.4 Array of Circular and Noncircular Holes

An array of microelectrodes is used to fabricate microhole patterns with circular or noncircular hole geometry. Chen (2008, 2007) used an array of microelectrodes and an upward batch EDM method to fabricate the rectangular hole pattern. Figure 4.9 shows a rectangular through-hole pattern fabricated using micro die-sinking EDM (Yi et al., 2008).

FIGURE 4.9
Through-hole pattern fabricated by batch-mode micro EDM (Yi et al., 2008).

4.4.5 Micromolds

Micro die-sinking EDM is widely used for fabrication of microdies for various applications (Uhlmann et al., 2005). Injection molds for mass production of microcomponents can be fabricated by die-sinking EDM.

4.5 Advantages

1. Die-sinking EDM does not require multiaxis CNC to perform. This makes the machine tool cheap and affordable.
2. In die-sinking EDM, the machining time required to fabricate noncircular holes and patterns is less compared to EDM milling.
3. As the tool does not rotate, no spindle motor and complex electrical connections are required.
4. As the tool moves only in a vertical direction, a less complex servo control mechanism can be used.
5. The reduced tool and worktable movements reduce the effect of vibration on the tool electrode.

4.6 Challenges

1. Die-sinking EDM requires additional tool fabrication techniques to make intrinsic shapes on the tool electrode.
2. On-machine fabrication of the tool electrode is challenging. Chance of tool handling-related errors arises if the tool is fabricated in a different machine tool.
3. The absence of tool rotation leads to inefficient dielectric flushing and debris accumulation.
4. Deposition of the workpiece material on the tool electrode is present due to debris melting and reattachment.
5. Percentage of shorting is observed to be higher in die-sinking EDM compared to other EDM variants.
6. To avoid short-circuiting and arcing, a reciprocating motion has to be provided. This nonproductive time increases the total machining time. Moreover, the vortex formed at the sides of the tool electrode during the reciprocating motion causes concave walls on the microholes (Murray et al., 2012).

4.7 Conclusion

Compared to other variants of micro EDM, die-sinking EDM offers a simple and economically viable manufacturing method. Die-sinking EDM can be used in the fabrication of fuel cells, micromixers, 3D microstructures, and arrayed noncircular holes. Dielectric flushing is challenging as the tool electrode has no rotation. Debris accumulation results in short-circuiting and arcing, which increases the machining time and deteriorates the machining quality. A layer of workpiece deposition is observed to be present on the tool surface due to remelting of debris particles.

References

Cao, M.R., Yang, S.Q., Li, W.H., Yang, S.C., 2010. Chip-ejection mechanism and experimental study of water dispersant dielectric fluid on small-hole EDM. *Adv. Mater. Res.* 97, 4111–4115.

Chen, S.T., 2007. A high-efficiency approach for fabricating mass micro holes by batch micro EDM. *J. Micromech. Microeng.* 17(10), 1961.

Chen, S.T., 2008. Fabrication of high-density micro holes by upward batch micro EDM. *J. Micromech. Microeng.* 18(8), 085002.

Hung, J.C., Chang, D.H., Chuang, Y., 2012. The fabrication of high-aspect-ratio micro-flow channels on metallic bipolar plates using die-sinking micro-electrical discharge machining. *J. Power Sources* 198, 158–163.

Liu, Y., Qu, Y., Zhang, W., Ma, F., Sha, Z., Wang, Y., Rolfe, B., Zhang, S., 2017. The effect of high frequency pulse on the discharge probability in micro EDM. *IOP Conference Series: Materials Science and Engineering*, Kunming, China, 281(1), 012031.

Murray, J., Zdebski, D., Clare, A.T., 2012. Workpiece debris deposition on tool electrodes and secondary discharge phenomena in micro-EDM. *J. Mater. Process. Technol.* 212, 1537–1547.

Schacht, B., Kruth, J.P., Lauwers, B., Vanherck, P., 2004. The skin-effect in ferromagnetic electrodes for wire-EDM. *Int. J. Adv. Manuf. Technol.* 23, 794–799.

Tong, H., Li, Y., Wang, Y., 2008. Experimental research on vibration assisted EDM of micro-structures with non-circular cross-section. *J. Mater. Process. Technol.* 208, 289–298.

Uhlmann, E., Piltz, S., Doll, U., 2005. Machining of micro/miniature dies and moulds by electrical discharge machining - recent development. *J. Mater. Process. Technol.* 167, 488–493.

Xu, B., Guo, K., Wu, X., Lei, J., Liang, X., Guo, D., Ma, J., Cheng, R., 2018a. Applying a foil queue micro-electrode in micro-EDM to fabricate a 3D micro-structure. *J. Micromech. Microeng.* 28, 55008.

Xu, B., Wu, X., Lei, J., Cheng, R., Ruan, S., Wang, Z., 2015. Laminated fabrication of 3D micro-electrode based on WEDM and thermal diffusion welding. *J. Mater. Process. Technol.* 221, 56–65.

Xu, B., Wu, X., Lei, J., Zhao, H., Liang, X., Cheng, R., Guo, D., 2018b. Elimination of 3D micro-electrode's step effect and applying it in micro-EDM. *Int. J. Adv. Manuf. Technol.* 96, 429–438.

Yi, S.M., Park, M.S., Lee, Y.S., Chu, C.N., 2008. Fabrication of a stainless steel shadow mask using batch mode micro-EDM. *Microsyst. Technol.* 14, 411–417. doi: 10.1007/s00542-007-0468-0.

5

Micro EDM Drilling

5.1 Introduction

Drilling is an essential step in manufacturing which enables assembling of microcomponents, connecting different layers in components, fixturing, and controlling flow and heat dissipation. In micromanufacturing, the challenge is to machine burr-free microholes with a minimum dimensional error. To drill outlet holes in micronozzles (Tong et al., 2013), drill starting hole for wire electro discharge machining (EDM) process, fabricate reverse EDM electrode (Mujumdar et al., 2010), and drill cooling holes in turbine blades (Kliuev et al., 2016) and inlet/outlet holes in microfluidic systems, the role of micro EDM drilling is inevitable. Imran et al. (2015) compared the characteristics of three microdrilling processes by drilling a microhole of 500 μm diameter and 2 mm length on Inconel alloy, and the details of the experimental analysis are given in Table 5.1. From this study, it is clear that micro EDM drilling has the advantages of smaller roundness error and thinner recast layer compared to the laser machining process. Compared to the conventional drilling method, micro EDM avoids the chance of tool breakage during drilling.

TABLE 5.1

Comparison of Different Processes for Microdrilling

Microhole Features	Microdrilling Processes		
	Micro EDM	Laser Drilling	Mechanical Drilling
Roundness error	7–9 μm	27–31 μm	6–7 μm
Metallurgical change to base material	Recast layer and heat affected zone (HAZ)	Nonuniform recast layer and HAZ	Fine grain structure layer and deformed layer
Hole edge profile	6–12 μm thick recast layer	9–45 μm thick recast layer	No recast layer. 3–6 μm thick, fine-grained zone
Observed defects	Radial microcracks	Radial microcracks	No cracks. Microburrs
Machining time (for the hole of 500 μm dia. and 2 mm length)	15 min	<1 s	~8 s

Source: Imran et al. (2015).

5.2 Principle of Operation

A rotating tool electrode is vertically advanced to the workpiece in an environment of continuous dielectric flow and pulsed DC supply, as shown in Figure 5.1. Then, the discharges produced during dielectric breakdown remove the material from the workpiece surface. The method of material removal remains the same (melting and vaporization) as in any other variants of micro EDM. However, compared to micro EDM milling, the tool bottom surface is responsible for most of the material removal action. Exempting some special cases for enhanced dielectric delivery, readily available cylindrical rods are commonly used as the tool electrode. To process the tools to exact dimensions, microwire EDG and microblock EDG are used. Similar to other variants of EDM, tool electrode wear is common during machining. To reduce the effect of tool erosion, appropriate wear compensation strategies have to be implemented.

As the tool progresses, only the entrance section of the hole will be open, and the removal of debris particle becomes challenging. As the drilling depth increases, ensuring the availability of dielectric fluid at the machining zone becomes more and more difficult. This results in a reduced percentage of normal discharges and increased percentage of harmful discharges or arcs. In a conventional EDM process, this challenge is tackled using tubular electrodes with internal flushing. However, as the size of the electrode reduces, the internal dielectric flushing becomes challenging, and high dielectric flow pressure may result in vibration or failure of the tool electrode. Furthermore, there is no significant improvement in surface quality with the use of tubular electrodes in micro EDM drilling (D'Urso and Merla, 2014). Many researchers have come up with some innovative techniques to

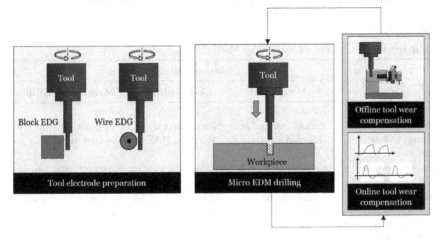

FIGURE 5.1
Schematic representation of different steps in micro EDM drilling.

FIGURE 5.2
Tool guide to ensure straightness and reduce tool vibrations during high-aspect-ratio drilling (Ferraris et al., 2013).

solve the problem of dielectric unavailability and debris removal inefficiency during deep-hole microdrilling. One of them is the use of orbital movement during micro EDM drilling (Bamberg and Heamawatanachai, 2008). A much smaller tool than the required microhole dimension and an orbital tool motion along with rotation and downward movement enhance dielectric flushing. The process is commonly known as planetary EDM drilling, which is considered to be an advanced variant of EDM.

For drilling deep holes using micro EDM, long slender tools have to be employed. These tools are susceptible to vibrations during machining due to short circuits, quick tool retrieval motion and shocks due to bubble explosions, etc. Moreover, as the length of the tool increases, the chance of tool deflection increases. To keep the tool straight and avoid vibrations, separate tool guide systems have to be attached to the spindle assembly as shown in Figure 5.2.

5.3 Machining Behavior

5.3.1 Effect of Debris Particles

The molten material washed away from the crater experiences sudden cooling and gets solidified to spherical globules. These debris particles are then rotated along with the stirring dielectric fluid. Compared to EDM milling, these particles cannot escape from the machining zone to the tank easily as they are surrounded by side and bottom walls of the hole.

However, in shallow holes, the dielectric jet can wash away the debris particles from the hole. As the depth of the hole increases, the dielectric fluid jet becomes inefficient in flushing out the debris particles. Figure 5.3 shows the forces acting on the debris particles during EDM drilling. F_B is the buoyancy force, F_D is the drag force, and F_G is the force due to debris weight. According to Haas et al. (2008), if the dielectric flow rate used in micro EDM is inadequate to evacuate the debris from the hole, it floats in the dielectric fluid ($F_G = F_B + F_D$). The presence of debris between the tool periphery and side wall of the hole introduces harmful discharges called secondary discharges, and the hole dimension will be severely affected (as shown in Figure 5.3). Similarly, the debris particles may merge and accumulate at the hole bottom affecting the effective spark gap. This may result in frequent short circuits and arc discharges. The effect of side sparking will be maximum at the hole entrance section because of the continuous action of sparks from the fresh surface as the tool descends.

5.3.2 Hole Depth

Similar to EDM milling, the tool experiences erosion of material from its surface. However, the side wear will be less compared to EDM milling, and in most of the cases, wear is limited to the tool bottom area. For through holes, tool wear will not be a severe problem as the sole aim is to drill through the complete thickness of the workpiece. For blind holes, the hole depth will be directly affected by tool wear, and drilling to the exact depth is a troublesome challenge to tackle. Special tool wear compensation strategies have to be implemented to solve this problem (Malayath et al., 2019a). The effect of hole depth and surface quality concerning various machining conditions during micro EDM drilling can be understood from Figure 5.4.

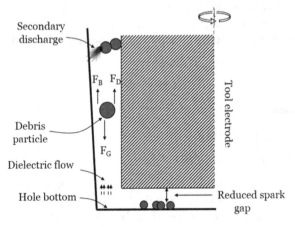

FIGURE 5.3
Machining behavior during micro EDM drilling.

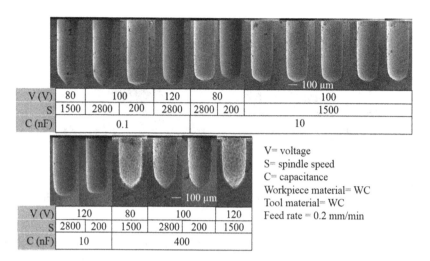

V (V)	80	100		120	80		100			
S	1500	2800	200	2800	2800	200	1500			
C (nF)	0.1				10					

V (V)	120		80	100		120
S	2800	200	1500	2800	200	1500
C (nF)	10		400			

V= voltage
S= spindle speed
C= capacitance
Workpiece material= WC
Tool material= WC
Feed rate = 0.2 mm/min

FIGURE 5.4
Effect of machining parameters on the surface quality and depth during micro EDM drilling (Malayath et al., 2019b).

Here, the tool travels to 1 mm depth, but the hole depth varies from 742 to 985 µm (Malayath et al., 2019b). As the discharge energy increases, the tool erosion becomes dominant, and the hole depth decreases.

5.3.3 Electrode Size

As the diameter of the electrode increases, the machining area increases. As a result, the parasitic capacitance or stray capacitance between the tool and the workpiece increases. This has a direct impact on the discharge energy. With the increase in stray capacitance, the discharge energy also increases. Moreover, the bigger electrodes make debris removal more difficult, which leads to an elevated percentage of abnormal discharges. Due to this "area effect," material removal rate (MRR) and tool wear will increase (Liu et al., 2016b). The taper angle of the hole increases with an increase in the machining area. This area effect also has a significant influence on the material migration phenomena during micro EDM milling (Wang et al., 2018).

5.3.4 Grain Size and Crystallographic Orientation of Workpiece Material

The machining behavior of a micro EDM drilling operation inside a grain and at the grain boundary of the workpiece differs considerably due to the difference in physical and chemical properties between these two locations (Liu et al., 2017). Figure 5.5 shows microholes drilled within the grain and across the grain boundary. The thermal conductivity and

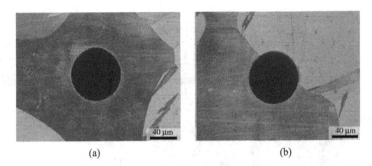

FIGURE 5.5
Micro EDM drilling (a) within a grain and (b) across a grain boundary (Li et al., 2013b).

melting point of the grain boundary are observed to be lower than that of the grain, which increases the MRR (Li et al., 2013b). Moreover, smaller grain size increases grain boundary volume fraction, which reduces the effective thermal conductivity. Lowering the thermal conductivity leads to easy removal of workpiece material. As the heat dissipation to the surrounding surface will be small in materials with less thermal conductivity, it will attain its melting point easily. Also, the shape and size of the discharge crater depend on the crystallographic orientation of the workpiece material (Liu et al., 2016a).

5.4 Process Capabilities and Applications

Pioneering research in micro EDM drilling was aimed to drill microholes with minimum possible dimensions to understand the process capability. Asad et al. (2007) used different tool fabrication methods to fabricate micro EDM electrodes. Then, microholes of diameter down to 6.5 µm are machined on a stainless steel plate of 50 µm (Figure 5.6) thickness to establish the applicability of EDM towards deep-hole microdrilling.

5.4.1 Deep-Hole Microdrilling

High-aspect-ratio microholes are used in optoelectronic instruments, micro electro-mechanical system (MEMS) devices, and drug delivery systems (Masuzawa et al., 1989). During conventional mechanical drilling, as the depth increases, the chance of tool breakage also increases. It is challenging to remove the broken drill bits from the microholes. Micro EDM drilling can be a viable choice considering the contact-free nature of the machining process. Masuzawa et al. (1989) developed a horizontal micro EDM setup to drill deep microholes, considering the possibility of gravity-assisted debris

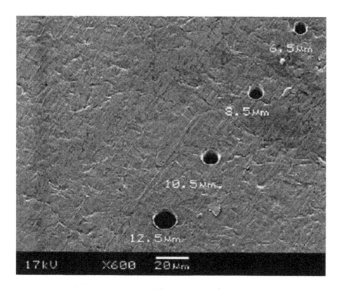

FIGURE 5.6
Microholes drilled with microelectrodes fabricated utilizing different tool fabrication methods (Asad et al., 2007).

removal. Microholes with an aspect ratio of 10 are machined with this unique strategy. Yu et al. (2002) drilled high-aspect-ratio microblind holes by providing planetary motion to the tool electrode. To reduce taper (difference in the entrance and exit diameter) during deep-hole microhole drilling, a varying capacitance method is employed by Kim et al. (2006). The capacitance value is changed during machining at specific depth intervals from 500 to 3,000 pF. The capability of micro EDM drilling to cut very hard materials including WC-Co and SUS 304 is conducted by Jahan et al. (2010). In this study, the thermal properties of the workpiece materials were found to affect the machining performance of micro EDM drilling significantly. WC-Co is the better choice as a workpiece material for burr-free microholes using EDM, regarding MRR, surface roughness, and less percentage of ineffective pulses. Figure 5.7 shows a high-aspect-ratio microhole drilled on stainless steel (SS 304) material with a discharge energy of 11 µJ.

Microholes with 200 µm diameter and an aspect ratio of 120 is machined by insulating the tool electrode with Parylene C polymer (Figure 5.8) (Ferraris et al., 2013). Jahan et al. (2012) fabricated microholes with an aspect ratio of 17 by employing low-frequency workpiece vibrations. However, ultrasonic vibrations may affect the shape accuracy and surface integrity during deep microhole drilling (Billa et al., 2010). When EDM became a popular method for deep-hole microdrilling, the need for predicting the achievable aspect ratio for an electrode–workpiece combination at different machining conditions became necessary.

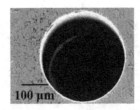

FIGURE 5.7
A high-aspect-ratio (AR = 7.5) microhole drilled on SUS 304 (Jahan et al., 2010).

FIGURE 5.8
High-aspect-ratio microholes machined with polymer-coated EDM tools (Ferraris et al., 2013).

Li et al. (2013a) introduced a theoretical model based on fluid mechanics, surface tension, and the dynamic contact angle to predict the maximum aspect ratio during the drilling process. According to the model, when the machining rate reaches zero, there will be a balance between the bubble pressure at the microhole bottom and outside pressure. Experimental validation of the model suggested its applicability in predicting the possible aspect ratio of microholes. However, the model did not consider the effect of discharge energy, debris accumulation, electrode rotation, and temperature gradients during machining. The usage of the mist nozzle is also reported to be improving the machining time and aspect ratio of microholes (Natsu and Maeda, 2018). The feasibility of using micro EDM drilling for machining deep microholes in ceramic materials (siliconized silicon carbide) is investigated by Skoczypiec et al. (2015). It is reported that the feed rate in ceramic microhole drilling using EDM is as high as 2 mm/min.

5.4.2 Drilling Microholes in Carbon Fiber-Reinforced Plastics

Due to better strength-to-weight ratio, carbon fiber-reinforced plastics (CFRPs) have been widely used in aerospace, biomedical, and automobile industries, and they are used also to make sports accessories. However, microhole drilling using conventional methods results in high tool wear, high delamination, and severe fiber pullouts. Micro EDM is used as an alternative to conventional drilling methods to drill microholes in CFRPs. Teicher et al. (2013) drilled microholes with EDM tool electrodes of 300 and 200 μm diameter. Park et al. (2015) conducted micro EDM drilling experiments on CFRPs to study the effects of various machining parameters on

FIGURE 5.9
Microhole drilled on CFRPs using electrodes of special tool geometry (Kumar et al., 2018).

the machining performance. Kumar et al. (2018) conducted an extensive analysis of high-aspect-ratio microdrilling operation (maximum aspect ratio = 29.17) of CFRP with EDM electrodes of various tool geometries (solid, single notch, and double notch). Figure 5.9 shows microholes drilled on CFRP by EDM.

5.4.3 Spray Holes in Fuel Jet Nozzles

To control the fuel volume and jet pressure, spray holes in fuel injectors must have a high dimensional accuracy. Moreover, the spray holes must have very small diameter (<300 µm), thick wall (>1 mm), high aspect ratio, and a taper angle in the range of 0°–2° (in the direction of the fuel jet) (Tong et al., 2012, 2013). Considering all these demands, micro EDM drilling can be the perfect choice for spray hole drilling in fuel injectors. A method to fabricate microholes with reverse taper using micro EDM is introduced by Diver et al. (2004). Spray holes of outlet diameter near to 250 µm are machined with accurate taper angle using a taper swinging mechanism (Tong et al., 2013) as shown in Figure 5.10. The taper angle is controlled within ±0.1°, and the dimensional accuracy is controlled within ±2 µm.

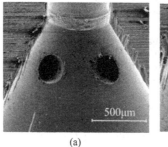
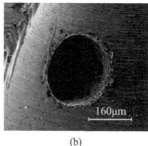

(a) (b)

FIGURE 5.10
(a) Micro spray holes for a fuel injector machined by EDM and (b) outlet of the spray hole (Tong et al., 2013).

5.4.4 Cooling Holes in Turbine Blades

High temperature inside the jet engines demands internal and external cooling of the turbine blades. Long cooling paths that are open to diffusor openings have to be drilled to cool the turbine internally. Inconel 718 is widely used as the material for turbine blades. Drilling microholes in hard-to-cut materials like Inconel is very difficult with conventional machining techniques. Micro EDM is successfully used to make microholes for cooling the turbine blades (Kliuev et al., 2016).

5.4.5 Biomedical Implants

Ti-6Al-4V is one of the widely used materials for biomedical implants. Micro EDM is successfully used to process Ti alloy (Plaza et al., 2014; Li et al., 2016; Tiwary et al., 2015; Sarhan et al., 2015).

5.4.6 Multifiber Titanium Ferrules

Ultrasmall holes are drilled using micro EDM drilling with high precision on multifiber optical connectors used in Mars rover (case study: one single process for multifiber titanium ferrule, www.sarix.com). Table 5.2 summarizes different research findings related to micro EDM drilling of different materials.

TABLE 5.2

Micro EDM Drilling of Different Materials

Ni Alloys	
Liu et al. (2005)	EDM drilling is used to fabricate microholes on nickel–molybdenum alloy for the applications in MEMS devices, particularly for enhancing magnetic shielding properties. A two-stepped electrode with multiple diameters is used to increase the circularity of microholes and improve surface properties.
Hung et al. (2007)	A two-step electrode with silicon carbide deposition improved the quality of the hole surface to a R_a=0.462 μm.
Ay et al. (2013)	Hole taper and overcut during micro EDM drilling of Inconel alloy is correlated with pulse duration and pulse current with the help of gray rational analysis. Changes in pulse duration and pulse current have a direct impact on the pulse energy, which influences the circularity and heat-affected zone in EDM drilling.
Imran et al. (2015)	Comparison of laser drilling, EDM drilling, and conventional drilling of microholes in Inconel 718 super alloy revealed the advantages and limitations of each process towards better dimensional control. EDM and laser proved the ability to drill deep microholes, but the surface quality is affected by the presence of recast layer depositions. Holes drilled by conventional microdrilling lacked small machining depth and the presence of microburrs on the surface.

(Continued)

TABLE 5.2 (*Continued*)

Micro EDM Drilling of Different Materials

Bhosle and Sharma (2017)	Optimization of the machining parameters is done with the help of gray rational analysis of micro EDM drilling on Inconel 600.
Perveen and Jahan (2018)	Drilling of microholes in Ni X alloy with tungsten carbide tool (normal and diamond coated) revealed the superiority of the coated tools in terms of overcut.
Feng et al. (2018)	With the help of a magnetic suspension system, the machining performance of micro EDM drilling is improved significantly. It showed a 30% increase in discharge percentage, a 11% decrease in arc discharge, and a 23% increase in MRR.
Ti Alloys	
Wansheng et al. (2002)	Effect of vibration-assisted EDM drilling process in improving the quality of microholes and maximum achievable drilling depth is investigated. Microholes with an aspect ratio of 15 is fabricated with the help of ultrasonic vibration.
Kibria et al. (2010)	Effect of different dielectric fluids and powder additives during EDM drilling of microholes is investigated. Powder additives enhance MRR during EDM drilling.
Meena and Azad (2012)	Microdrilling of Ti alloy (Ti-6Al-4V) is done to study the effect of machining parameters on the machining performance gray rational analysis. Voltage, current, pulse width, and frequency are recognized as the influential machining parameters in the respective order.
Plaza et al. (2014)	Feasibility of using helical microtools for better debris removal during EDM drilling is proved by experimental studies in Ti-6Al-4V. The effect of helix angle and flute length on MRR is studied.
Mondol et al. (2015)	A full factorial analysis of micro EDM drilling of Ti alloy suggested capacitance as the most influential machining parameter in controlling MRR, tool wear rate (TWR), and overcut.
Li et al. (2016)	EDM microdrilling on Ti alloy revealed the negative effect of low thermal conductivity of the workpiece material on the quality of the machined surface. Comparative study on stainless steel workpiece suggested that Ti alloys exhibit larger heat-affected zones.
Krishnaraj (2016)	A parametric study based on Taguchi design showed that peak current is the most critical parameter determining the performance of micro EDM drilling of Ti alloys.
Tiwary et al. (2015) and Alavi and Jahan (2017)	A multioptimization analysis of the quality control parameters of micro EDM drilling with the help of response surface methodology suggested discharge current, pulse on time, spark gap as well as flushing pressure as the prominent factors determining the MRR, TWR, overcut, and taper.
Tungsten Carbide	
Malayath et al. (2019b)	A parametric analysis of the tungsten carbide microend mill fabrication using micro EDM is carried out. Microcutting tools of 100 µm diameter are successfully fabricated and tested the performance by milling microchannels on different materials.

(Continued)

TABLE 5.2 *(Continued)*

Micro EDM Drilling of Different Materials

Jahan et al. (2010) and D'Urso et al. (2016)	Drilling microholes on stainless steel and tungsten carbide material showed that electrical resistivity, thermal conductivity, specific heat capacity, thermal expansion coefficient, and melting point of the electrode as well as workpiece material have a significant effect on the machining time and machining accuracy. Deep-hole drilling of stainless steel is less stable compared to WC due to the increased percentage of ineffective pulses. Machinability of steel is limited by low thermal conductivity, elevated work hardening tendency, and high ductility compared to WC.
Pilligrin et al. (2017)	Gray rational analysis of micro EDM drilling experiments suggested that capacitance is the most prominent control parameter of machining performance followed by polarity, speed, and voltage.
Ekmekci (2009)	Analyzing the holes drilled by micro EDM showed the presence of microcracks on the surface even though low-energy pulses were used for material removal. This phenomenon is attributed to the formation of high contraction stress due to solidification of molten material thrown out from the crater on to the existing recast layer. Once the stress exceeds the ultimate strength, microcracks appear on the surface. The possibility of electrical discharges between the debris and tool electrode during blind hole drilling is also discussed.
Song et al. (2009)	The phenomenon of electrolytic corrosion during micro EDM drilling is discussed. To avoid the detrimental effects of electrolytic corrosion, a bipolar pulse generator and a triangular tool are suggested. As the electrochemical reaction prevails on the positively charged electrode, the bipolar pulse generator alters the polarity periodically and reduces the chance of electrolytic corrosion. Less surface area of the triangular tool as compared to a cylindrical tool reduces electrolytic corrosion.
Jahan et al. (2009)	A comparative study of micro EDM drilling using different pulse generators is conducted on WC material. The study showed the presence of burr-like recast layer on the surfaces machined with transistor-type pulse generator.
Stainless Steel	
Yahagi et al. (2012)	Micro EDM drilling with high-speed spindle system and electrostatic induction feeding is introduced. The tool wear ratio at 50,000 rpm is reported to be half of the value at 1,000 rpm.
Manivannan and Kumar (2017)	Cryogenically cooled micro EDM drilling process exhibits a 45% improvement in MRR and 46% reduction in overcut.
Li and Bai (2018)	A simulation study and experimental analysis on the effect side gap on debris removal, MRR, and tool wear for deep-hole micro EDM drilling are conducted. The improvement in the performance parameters suggests that the alternating side gap method has a commendable advantage in employing for deep-hole drilling.
Other Materials	
Liew et al. (2014) Silicon carbide	Very small holes (~10 μm) with an aspect ratio as high as 420 is machined on SiC material. The advantages of using ultrasonic vibration to the workpiece and adding nanofibers to the dielectric are also discussed.
Somashekhar et al. (2010) Aluminum	The feasibility of using a neural network for optimizing the machining parameters is discussed. The proposed method exhibited a prediction error of 0.8312% during training and 3.94% during actual experiments.

5.5 Advantages

- Complex-shaped drilling tools are not required.
- No flutes or complex cutting geometry is required for material removal.
- Use of simple-shaped tools reduces tool fabrication time.
- No burr formation.
- Force-free operation. Tool breakage during drilling can be avoided.
- Demands simple fixtures.
- Microholes can be machined easily on difficult-to-cut materials.
- High-aspect-ratio microholes can be machined without tool deflection.

5.6 Challenges

- The entrance diameter enlargement due to side sparking is common.
- Blind hole drilling with depth accuracy is difficult to achieve due to tool erosion.
- As the depth of the hole increases, the MRR decreases severely.
- The percentage of harmful discharges is high due to the inefficiency in debris removal.
- For deep microhole drilling, the slender tool with large tool length will result in tool vibrations and a further increase in hole diameter.
- Innovative methods for tool wear control, including the use of insulated tools, increase the cost of machining and total machining time.

5.7 Conclusion

Micro EDM drilling is one of the prominent machining operations to fabricate accurate microholes. This method ensures a burr-free edge for the microhole without any special cutting tools or complex tool geometry. Application area includes aerospace industry (cooling holes on turbine blades), automobile industry (spray holes for fuel injection nozzles), biomedical implants, space exploration accessories, etc. To increase the surface integrity, special tool wear compensation techniques, tool electrode guides, optimum drilling conditions, and appropriate tool materials have to be selected.

References

Alavi, F., Jahan, M.P., 2017. Optimization of process parameters in micro-EDM of Ti-6Al-4V based on full factorial design. *Int. J. Adv. Manuf. Technol.* 92, 167–187.

Asad, A.B.M.A., Masaki, T., Rahman, M., Lim, H.S., Wong, Y.S., 2007. Tool-based micro-machining. *J. Mater. Process. Technol.* 192–193, 204–211. doi: 10.1016/j.jmatprotec.2007.04.038.

Ay, M., Çaydaş, U., Hasçalık, A., 2013. Optimization of micro-EDM drilling of Inconel 718 superalloy. *Int. J. Adv. Manuf. Technol.* 66, 1015–1023.

Bamberg, E., Heamawatanachai, S., 2008. Orbital electrode actuation to improve efficiency of drilling micro-holes by micro-EDM. *J. Mater. Process. Technol.* 9, 1826–1834. doi: 10.1016/j.jmatprotec.2008.04.044.

Bhosle, R.B., Sharma, S.B., 2017. Multi-performance optimization of micro-EDM drilling process of Inconel 600 alloy. *Mater. Today Proc.* 4, 1988–1997.

Billa, S., Sundaram, M.M., Rajurkar, K.P., 2010. A study on the high aspect ratio micro hole drilling using ultrasonic assisted micro electro discharge machining. *Proceedings of the ASPE 2007 Spring Topical Meeting*, Chapel Hill, NC, 32–36.

D'Urso, G., Maccarini, G., Ravasio, C., 2016. Influence of electrode material in micro-EDM drilling of stainless steel and tungsten carbide. *Int. J. Adv. Manuf. Technol.* 85, 2013–2025. doi: 10.1007/s00170-015-7010-9.

D'Urso, G., Merla, C., 2014. Workpiece and electrode influence on micro-EDM drilling performance. *Precis. Eng.* 38, 903–914. doi: 10.1016/j.precisioneng.2014.05.007.

Diver, C., Atkinson, J., Helml, H.J., Li, L., 2004. Micro-EDM drilling of tapered holes for industrial applications. *J. Mater. Process. Technol.* 149, 296–303. doi: 10.1016/j.jmatprotec.2003.10.064.

Ekmekci, B., 2009. Geometry and surface damage in micro electrical discharge machining of micro-holes. *J. Micromech. Microeng.* 19. doi: 10.1088/0960-1317/19/10/105030.

Feng, Y., Guo, Y., Ling, Z., Zhang, X., 2018. Micro-holes EDM of superalloy Inconel 718 based on a magnetic suspension spindle system.

Ferraris, E., Castiglioni, V., Ceyssens, F., Annoni, M., Lauwers, B., Reynaerts, D., 2013. EDM drilling of ultra-high aspect ratio micro holes with insulated tools. *CIRP Ann. Manuf. Technol.* 62, 191–194. doi: 10.1016/j.cirp.2013.03.115.

Haas, R., Munz, M., Huber, M., Knabe, R., 2008. Adequate gap flushing in high speed EDM drilling of deep small holes in moulds and dies. *Seventh International Conference on High Speed Machining*, Xi'an, China.

Hung, J.-C., Wu, W.-C., Yan, B.-H., Huang, F.-Y., Wu, K.-L., 2007. Fabrication of a micro-tool in micro-EDM combined with co-deposited Ni–SiC composites for micro-hole machining. *J. Micromech. Microeng.* 17, 763.

Imran, M., Mativenga, P.T., Gholinia, A., Withers, P.J., 2015. Assessment of surface integrity of Ni superalloy after electrical-discharge, laser and mechanical micro-drilling processes. *Int. J. Adv. Manuf. Technol.* 79, 1303–1311. doi: 10.1007/s00170-015-6909-5.

Jahan, M.P., San Wong, Y., Rahman, M., 2010. A comparative experimental investigation of deep-hole micro-EDM drilling capability for cemented carbide (WC-Co) against austenitic stainless steel (SUS 304). *Int. J. Adv. Manuf. Technol.* 46, 1145–1160.

Jahan, M.P., Wong, Y.S., Rahman, M., 2009. A study on the quality micro-hole machining of tungsten carbide by micro-EDM process using transistor and RC-type pulse generator. *J. Mater. Process. Technol.* 209, 1706–1716. doi: 10.1016/j.jmatprotec.2008.04.029.

Jahan, M.P., Wong, Y.S., Rahman, M., 2012. Evaluation of the effectiveness of low frequency workpiece vibration in deep-hole micro-EDM drilling of tungsten carbide. *J. Manuf. Process.* 14, 343–359. doi: 10.1016/j.jmapro.2012.07.001.

Kibria, G., Sarkar, B.R., Pradhan, B.B., Bhattacharyya, B., 2010. Comparative study of different dielectrics for micro-EDM performance during microhole machining of Ti-6Al-4V alloy. *Int. J. Adv. Manuf. Technol.* 48, 557–570.

Kim, D.J., Yi, S.M., Lee, Y.S., 2006. Straight hole micro EDM with a cylindrical tool using a variable capacitance method accompanied by ultrasonic vibration. *J. Micromech. Microeng.* 16. doi: 10.1088/0960-1317/16/5/031.

Kliuev, M., Boccadoro, M., Perez, R., Bó, W.D., Stirnimann, J., Kuster, F., 2016. EDM drilling and shaping of cooling holes in Inconel 718 turbine blades. *Procedia CIRP* 42, 322–327. doi: 10.1016/j.procir.2016.02.293.

Krishnaraj, V., 2016. Optimization of process parameters in micro-EDM of Ti-6Al-4V alloy. *J. Manuf. Sci. Prod.* 16, 41–49.

Kumar, R., Kumar, A., Singh, I., 2018. Electric discharge drilling of micro holes in CFRP laminates. *J. Mater. Process. Technol.* 259, 150–158. doi: 10.1016/j.jmatprotec.2018.04.031.

Li, J., Yin, G., Wang, C., Guo, X., Yu, Z., 2013a. Prediction of aspect ratio of a micro hole drilled by EDM. *J. Mech. Sci. Technol.* 27, 185–190. doi: 10.1007/s12206-012-1214-9.

Li, J.Z., Shen, F.H., Yu, Z.Y., Natsu, W., 2013b. Influence of microstructure of alloy on the machining performance of micro EDM. *Surf. Coat. Technol.* 228, S460–S465.

Li, Y., Yu, W., Xu, J., Yu, H., 2016. A comparative investigation of micro-EDM drilling capability for Ti-6Al-4V alloy against austenitic stainless steel SUS 316. *2016 IEEE International Conference on Mechatronics and Automation (ICMA)*, Harbin, China. IEEE, 1612–1616.

Li, Z., Bai, J., 2018. Influence of alternating side gap on micro-hole machining performances in micro-EDM. *Int. J. Adv. Manuf. Technol.* 94, 979–989. doi: 10.1007/s00170-017-0959-9.

Liew, P.J., Yan, J., Kuriyagawa, T., 2014. Fabrication of deep micro-holes in reaction-bonded SiC by ultrasonic cavitation assisted micro-EDM. *Int. J. Mach. Tools Manuf.* 76, 13–20. doi: 10.1016/j.ijmachtools.2013.09.010.

Liu, H.-S., Yan, B.-H., Huang, F.-Y., Qiu, K.-H., 2005. A study on the characterization of high nickel alloy micro-holes using micro-EDM and their applications. *J. Mater. Process. Technol.* 169, 418–426.

Liu, Q., Zhang, Q., Zhang, M., Zhang, J., 2016a. Review of size effects in micro electrical discharge machining. *Precis. Eng.* 44, 29–40. doi: 10.1016/j.precisioneng.2016.01.006.

Liu, Q., Zhang, Q., Zhang, M., Zhang, J., 2017. Effects of grain size of AISI 304 on the machining performances in micro electrical discharge machining. *Proc. Inst. Mech. Eng. Part B J. Eng. Manuf.* 231, 359–366.

Liu, Q., Zhang, Q., Zhu, G., Wang, K., Zhang, J., Liu, Q., Zhang, Q., Zhu, G., Wang, K., Zhang, J., Dong, C., 2016b. Effect of electrode size on the performances of effect of electrode size on the performances of micro-EDM. *LMMP* 31, 391–396. doi: 10.1080/10426914.2015.1059448.

Malayath, G., Katta, S., Sidpara, A.M., Deb, S., 2019a. Length-wise tool wear compensation for micro electric discharge drilling of blind holes. *Meas. J. Int. Meas. Confed.* 134. doi: 10.1016/j.measurement.2018.12.047.

Malayath, G., Sidpara, A.M., Deb, S., 2019b. Fabrication of micro end mill tool by EDM and its performance evaluation. *J. Mach. Sci. Technol.* (Accepted).

Manivannan, R., Kumar, M.P., 2017. Multi-attribute decision-making of cryogenically cooled micro-EDM drilling process parameters using TOPSIS method. *Mater. Manuf. Process.* 32, 209–215. doi: 10.1080/10426914.2016.1176182.

Masuzawa, T., Tsukamoto, J., Fujino, M. (1989). Drilling of deep microholes by EDM. *CIRP Annals*, 38(1), 195–198.

Meena, V.K., Azad, M.S., 2012. Grey relational analysis of micro-EDM machining of Ti-6Al-4V alloy. *Mater. Manuf. Process.* 27, 973–977.

Mondol, K., Azad, M.S., Puri, A.B., 2015. Analysis of micro-electrical discharge drilling characteristics in a thin plate of Ti – 6Al – 4 V 141–150. *Int. J. Adv. Manuf. Technol.* 76, 141–150. doi: 10.1007/s00170-013-5414-y.

Mujumdar, S.S., Mastud, S.A., Singh, R.K., Joshi, S.S., 2010. Experimental characterization of the reverse micro-electrodischarge machining process for fabrication of high-aspect-ratio micro-rod arrays. *Proc. Inst. Mech. Eng. Part B J. Eng. Manuf.* 224, 777–794. doi: 10.1243/09544054JEM1745.

Natsu, W., Maeda, H., 2018. Realization of high-speed micro EDM for high-aspect-ratio micro hole with mist nozzle. *Procedia CIRP* 68, 575–577. doi: 10.1016/j.procir.2017.12.116.

Park, S., Kim, G., Lee, W., Min, B., Lee, S., Kim, T., 2015. Microhole machining on precision CFRP components using electrical discharging machining. *20th International Conference on Composite Materials*, Copenhagen, Denmark.

Perveen, A., Jahan, M.P., 2018. Comparative micro-EDM studies on Ni based X-alloy using coated and uncoated tools. *Mater. Sci. Forum.* 911, 13–19.

Pilligrin, J.C., Asokan, P., Jerald, J., Kanagaraj, G., Nilakantan, J.M., Pilligrin, J.C., Asokan, P., Jerald, J., Kanagaraj, G., Nilakantan, J.M., 2017. Tool speed and polarity effects in micro-EDM drilling of 316L stainless steel. *Prod. Manuf. Res.* 5, 99–117. doi: 10.1080/21693277.2017.1357055.

Plaza, S., Sanchez, J.A., Perez, E., Gil, R., Izquierdo, B., Ortega, N., Pombo, I., 2014. Experimental study on micro EDM-drilling of Ti-6Al-4V using helical electrode. *Precis. Eng.* 38, 821–827.

Sarhan, A.A.D., Fen, L.S., Yip, M.W., Sayuti, M., 2015. Fuzzy modeling for micro EDM parameters optimization in drilling of biomedical implants Ti-6Al-4V alloy for higher machining performance. *Int. J. Mech. Aerosp. Ind. Mech. Eng.* 9(1), 197–201.

Skoczypiec, S., Machno, M., Bizoń, W., 2015. The capabilities of electrodischarge microdrilling of high aspect ratio holes in ceramic materials. *Manage. Prod. Eng. Rev.* 6, 61–69. doi: 10.1515/mper-2015-0027.

Somashekhar, K.P., Ramachandran, N., Mathew, J., Ramachandran, N., Optimization, J.M., Somashekhar, K.P., Ramachandran, N., Mathew, J., 2010. Optimization of material removal rate in micro- EDM using artificial neural network and genetic algorithms. doi: 10.1080/10426910903365760.

Song, K.Y., Chung, D.K., Park, M.S., 2009. Micro electrical discharge drilling of tungsten carbide using deionized water. *Mater. Manuf. Processes* 25, 465–475. doi: 10.1088/0960-1317/19/4/045006.

Teicher, U., Müller, S., Münzner, J., Nestler, A., 2013. Micro-EDM of carbon fibre-reinforced plastics. *Procedia Cirp* 6, 320–325.

Tiwary, A.P., Pradhan, B.B., Bhattacharyya, B., 2015. Study on the influence of micro-EDM process parameters during machining of Ti–6Al–4V superalloy. *Int. J. Adv. Manuf. Technol.* 76, 151–160.

Tong, H., Li, Y., Zhang, L., 2012. Swing mechanism for micro EDM drilling of fuel jet nozzles. *Adv. Mater. Res.* 593, 391–395. doi: 10.4028/www.scientific.net/AMR.591-593.391.

Tong, H., Li, Y., Zhang, L., Li, B., 2013. Mechanism design and process control of micro EDM for drilling spray holes of diesel injector nozzles. *Precis. Eng.* 37, 213–221. doi: 10.1016/j.precisioneng.2012.09.004.

Wang, K., Zhang, Q., Zhu, G., Zhang, J., Wang, K., 2018. Effects of tool electrode size on surface characteristics in micro-EDM. *Int. J. Adv. Manuf. Technol.* 96, 3909–3916.

Wansheng, Z., Zhenlong, W., Shichun, D., Guanxin, C., Hongyu, W., 2002. Ultrasonic and electric discharge machining to deep and small hole on titanium alloy. *J. Mater. Process. Technol.* 120, 101–106.

Yahagi, Y., Koyano, T., Kunieda, M., Yang, X., 2012. Micro drilling EDM with high rotation speed of tool electrode using the electrostatic induction feeding method. *Procedia CIRP* 1, 162–165. doi: 10.1016/j.procir.2012.04.028.

Yu, Z.Y., Rajurkar, K.P., Shen, H., 2002. High aspect ratio and complex shaped blind micro holes by micro EDM. *CIRP Ann. Manuf. Technol.* 51, 359–362. doi: 10.1016/S0007-8506(07)61536-4.

6

Other Relevant Process Variants of Micro EDM

6.1 Introduction

The principle of material ablation using electrical discharges is used in electrodischarge grinding (EDG), planetary electro discharge machining (EDM), and reverse EDM (R-EDM). Micro EDM can be combined with other machining techniques to get better material removal rate (MRR) and surface quality. These special variants of micro EDM have a broad area of applications.

6.2 Electrodischarge Grinding

EDG is widely used for tool electrode preparation in micro EDM. Sacrificial electrodes of different shapes (wire, block, and disc) are used to shape the tool electrode for subsequent EDM milling or drilling.

6.2.1 Block EDG

This variant of micro EDM has a block of high wear resistance and melting point as a sacrificial electrode (negative polarity). Compared to the usual arrangement of micro EDM, the block is fixed in the worktable, and the workpiece rod (which may be used as the tool electrode for subsequent machining operation) is attached to the spindle (positive polarity). The rod is then moved along the block surface to get the intended tool dimensions. Zhao et al. (2006) conducted a parametric study of block EDG process and reported that the tangential feed during EDG could be helpful to improve the dimensional accuracy of microrods. Micro EDG is categorized into two according to the relative motion between the block and the rod.

6.2.1.1 Stationary EDG

Here, the block remains stationary, and the rod moves down into the block as shown in Figure 6.1. The final tool length is decided by the vertical feed, and the final tool diameter is decided by the overlap between the rod and the stationary block. During machining, the block experiences a continuous erosion of material along the direction of motion of the electrode. Due to the wear of the block, the final rod may have a nonuniform shape and a larger diameter than expected. However, only a small area of the face of the block will be used for machining, and the block can be used for a more extended period.

6.2.1.2 Moving Block EDG

Here, the block also moves to create a simultaneous two-direction motion along a face of the block (Figure 6.2), which reduces the errors due to material erosion from the sacrificial block. Compared to the stationary block, this combined motion ensures a fresh block surface as machining progresses and gives much more control over the final tool dimensions. However, as the block uses more area for processing the rod, the block erodes quickly and demands frequent replacement or regrinding of the sacrificial electrode.

6.2.2 Wire EDG

Masuzawa et al. (1985) introduced the method of using a continuously running wire (negative polarity) to shape the electrode (positive polarity) by which the effect of tool wear is reduced to a minimum (Figure 6.3). As the machining area is less, the time taken for tool fabrication will be higher.

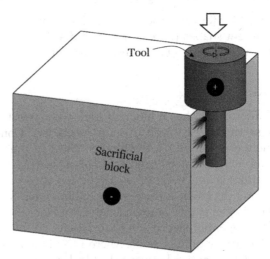

FIGURE 6.1
Schematic representation of a stationary block EDG arrangement.

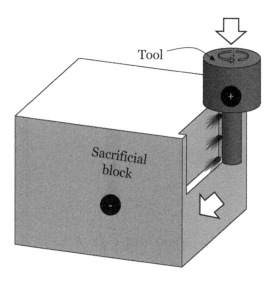

FIGURE 6.2
Schematic representation of moving block EDG arrangement.

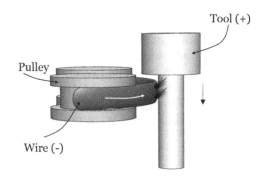

FIGURE 6.3
Schematic representation of wire EDG arrangement.

Moreover, the wire cannot be reused and has to be changed from time to time. However, wire EDG helps to achieve very small tool dimensions with the help of ultrasmall wire electrodes.

6.2.3 Disc EDG

In disc-type EDG, the only change is in the shape of the sacrificial electrode. A disc attached to the worktable and rotating in the vertical axis is used to shape the electrode with the help of appropriate tool motions as shown in Figure 6.4. Compared to other varieties of EDG, the disc type is capable of controlling the tool dimensions more accurately (Rahman et al., 2007).

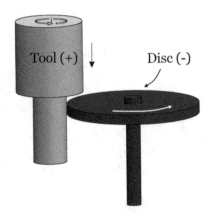

FIGURE 6.4
Schematic representation of disc-type EDG arrangement.

6.2.4 Applications

Tool electrode preparation: Micro EDG is mainly used for on-machine fabrication of the tool electrode. Shaping the tool electrode to intended dimensions soon after fixing the electrode to the spindle reduces the chance of clamping errors and tool handling-related errors. Asad et al. (2007) discussed various on-machine tool fabrication methods and presented EDG as one of the most efficient methods to shape the tool electrode with minimum possible errors. Figure 6.5 shows the difference in tool electrodes shaped by different types of EDG process. Tool electrodes processed by EDG are used to drill microholes on various materials (Ravi and Huang, 2002; Qingfeng et al., 2016).

6.2.4.1 Microgrinding Tool

A microgrinding tool for finishing of glass materials (BK7, Lithosil, and N-SF14) is fabricated using block EDG (Perveen et al., 2012). Figure 6.6 shows a combined setup for microgrinding tool fabrication and grinding. Surface

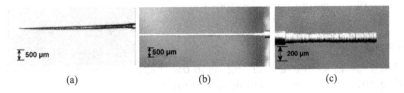

(a)   (b)   (c)

FIGURE 6.5
Tool electrodes fabricated by (a) stationary block EDG, (b) disc-type EDG, and (c) wire EDG (Rahman et al., 2007).

| Spindle head |
| PCD tool with tool with tool holder |
| Block micro EDM set-up |
| Micro EDM set-up with separate tank |
| Machine table containing both micro EDM and grinding in one step |

| Glass materials |
| Nozzle for dielectric and coolant |
| Micro grinding set-up |
| Valve for controlling coolant pressure |

FIGURE 6.6
A setup for on-machine fabrication of microgrinding tools using EDG (Perveen et al., 2012).

roughness as minimum as 12.79 nm is achieved through grinding with tools fabricated by block EDG. Another miniature grinding tool is fabricated by wire EDG and precision codeposition of diamond (Chen et al., 2008; Hung et al., 2007). A polycrystalline diamond (PCD) spherical grinding tool is fabricated by wire EDG and used to make concave surfaces on silicon and a convex surface on alumina ceramics (Masaki et al., 2007). Very high surface quality ($R_a = 5$ nm) is obtained while finishing with the spherical PCD tool.

6.2.4.2 Micromilling Tools

Micromilling tools of various tool geometries are fabricated using wire EDG by Morgan et al. (2006). With the help of those microcutting tools, microchannels are machined on aluminum with a surface roughness of 121 nm and on glass with a surface roughness of 5.7 nm. Conical and cylindrical micromilling tools are fabricated with the help of wire EDG and sacrificial block EDG (Morgan et al., 2004). Zhang et al. (2013) used wire EDG to fabricate hemispherical PCD microend mill tool as shown in Figure 6.7a. Chern et al. (2007), Ali and Ong (2006), Oliaei et al. (2013), and Zhan et al. (2015) also employed EDG to fabricate end mill tools.

6.2.4.3 Punching Tools

Wire EDG is used to fabricate noncircular microelectrodes to be used as punches in microforming operation (Chern and de Wang, 2007). A series of microholes of triangular, rectangular, and hexagonal shapes are fabricated with micropunches of respective geometry fabricated using EDG. A cylindrical micropunch fabricated using EDG is shown in Figure 6.7b.

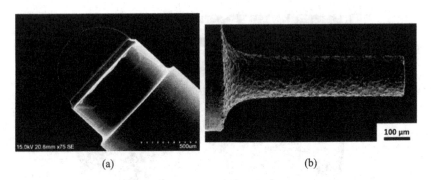

FIGURE 6.7
(a) Hemispherical microend mill tool (Zhang et al., 2013) and (b) micropunch fabricated by EDG (Xu et al., 2014).

6.3 Planetary Micro EDM

Planetary micro EDM or orbital micro EDM refers to the variant of EDM in which a planetary motion is provided for the tool electrode to fabricate microholes. The machining strategy for planetary EDM drilling is explained in Figure 6.8. The microelectrode is fed into the workpiece with orbital motion, and the material is removed with the help of different tool paths. Planetary EDM is primarily developed to eradicate the difficulty in flushing during high-aspect-ratio microdrilling operations (Yu et al., 2002). The orbital motion ensures adequate dielectric flow in the machining region to flush the debris out from the machining zone. Reducing the debris concentration in the machining zone will result in a reduced number of undesirable discharges, and the tool bottom wear can be controlled.

Egashira et al. (2006) conducted an experimental analysis of the planetary EDM process and revealed the advantageous effect of using orbital

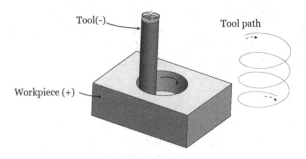

FIGURE 6.8
Planetary EDM drilling strategy.

motion during EDM drilling. With the help of planetary motion, the MRR was improved, along with a reduction in volumetric wear and overcut (<1 μm) (Egashira et al., 2006). Moreover, a microhole with an aspect ratio of 29 is drilled on a stainless steel workpiece with the aid of planetary movement using a horizontal micro EDM arrangement. A theoretical model of the planetary EDM is developed by Guo et al. (2013) to determine the optimum velocity of the planetary movement for better machining performance. A relationship connecting the machining rate, the cross-sectional area of the hole, the radius of the planetary movement, and the velocity of planetary movement is developed from the experimental analysis. Three types of orbital actuation during micro EDM drilling are shown in Figure 6.9 (Bamberg and Heamawatanachai, 2008). In the first method (Figure 6.9a), the hole is divided into specific cylindrical layers of thickness t (Z-slicing),

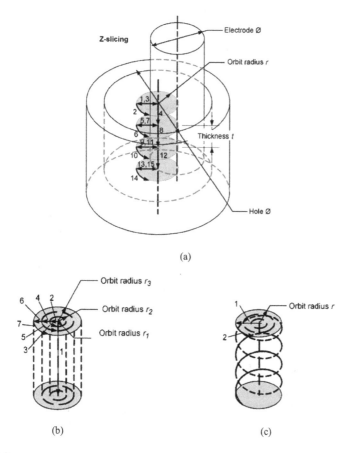

(a)

(b)

(c)

FIGURE 6.9
Different orbiting strategies for EDM electrode: (a) Z-slicing, (b) cylindrical slicing, and (c) helical tool path (Bamberg and Heamawatanachai, 2008).

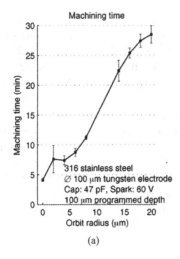
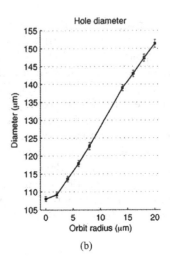

FIGURE 6.10

Effect of the orbital radius on (a) machining time and (b) final hole diameter during planetary micro EDM (Bamberg and Heamawatanachai, 2008).

and the material in each layer is removed by a circular sweep with the orbit radius of r. In the second method (Figure 6.9b), the tool electrode is plunged into the workpiece to the predetermined hole depth, and the orbital radius is increased gradually to expand the hole to the designed diameter. In the final strategy (Figure 6.9c), the tool is allowed to move in a spiral tool path. Bamberg and Heamawatanachai (2008) employed the Z-slicing strategy with a Profile 24P control system and a dedicated control algorithm (for pulling the electrode towards the center in the case of short circuit) for drilling microholes. The study revealed the effect of orbital radius on different performance parameters, which is shown in Figure 6.10. As the orbital radius increases, the machining time (Figure 6.10a) and overcut (Figure 6.10b) also increase.

6.4 Reverse Micro EDM

Normally, EDM tool electrodes are rods, wires, or blocks. However, in some cases, a plate with single or multiple through holes can be used as an EDM tool electrode. Similar to the extrusion process, as the EDM tool electrode with holes is plunged to the workpiece, some micropillars appear on the workpiece due to material erosion as explained in Figure 6.11a. The diameter of the rod depends on the diameter of the hole on the EDM plate electrode, and the shape of the rod will be the reverse replica of the hole shape.

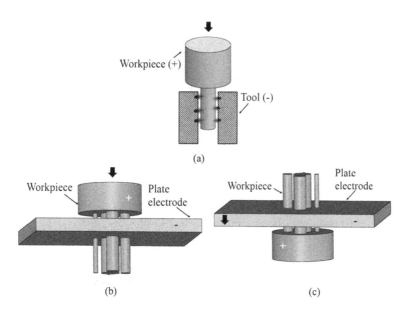

FIGURE 6.11
(a) R-EDM strategy, (b) normal R-EDM, and (c) inverse R-EDM.

This variant of micro EDM is popularly known as R-EDM (Mastud et al., 2012c). The type of R-EDM where the plate electrode remains stationary and the workpiece is plunged is known as normal R-EDM (Figure 6.11b). When the motion is given to the plate electrode instead of the workpiece electrode, it is called inverse R-EDM (Figure 6.11c) (Mastud et al., 2012b). While fabricating multiple electrodes, the plate electrode or the workpiece cannot be rotated. The absence of rotation and the presence of large machining area make the control of machining conditions in R-EDM more challenging.

Compared to other variants of micro EDM, debris entrapment is a bigger issue in R-EDM. This is because of the existence of larger machining zone compared to other processes. A sudden increase in the debris concentration at the primary machining zone happens after initiation of the machining process. These debris particles are then transported to the secondary region inside the hole, as shown in Figure 6.12 (Mastud et al., 2011). Entrapment of debris particle between the pillar side surface and the hole leads to welding of the microglobules to the pillar surface. Finally, the surface quality of the fabricated pillar is adversely affected by the presence of welded debris particles.

Hwang et al. (2010) presented three methods for enhancing the debris removal, namely, working fluid spraying, vibration-assisted electrode, and using shake-down-type workpiece. R-EDM is the easiest way to fabricate an array of micropillars on any conducting workpiece surface. Micropillar array can be used as micro electro chemical machining (ECM) tool electrode (Wang and Zhu, 2009), micro EDM tool electrode for microslitting

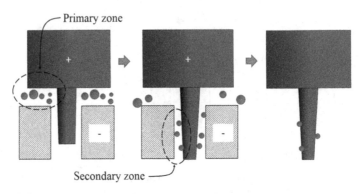

FIGURE 6.12
Welding of debris particles to the micropillars during R-EDM.

(Fleischer et al., 2004), neural signal capturing devices (Weiliang et al., 2006), biomedical applications (Penache et al., 2002), etc. However, the challenges in employing R-EDM for micromachining applications include low process stability, deflections of microrods due to thermal load, large recast layers, high surface roughness, and an elevated chance of surface burning.

6.4.1 Applications and Process Capabilities

- Mastud et al. (2012b) fabricated 10×10 arrays of WC microelectrodes of diameter less than $100\,\mu m$ with inverse R-EDM using W-Cu electrode of $300\,\mu m$ thickness.
- Hwang et al. (2010) fabricated a micropin array having a 40×40 arrangement and $30\,\mu m$ average diameter using $600\,\mu m$ thick brass plate (Figure 6.13a).
- Yi et al. (2008) used AISI 304 foil (Figure 6.13b) to fabricate arrayed microelectrodes (3×4) on the Cu electrode (Figure 6.13c).
- Weiliang et al. (2006) fabricated a $10\,\mu m$ 3×3 array of microelectrodes with an aspect ratio of 10 on AgW.
- Kim et al. (2006a) fabricated three high-aspect-ratio micropillars of $35\,\mu m$ diameter on a WC rod of $300\,\mu m$ diameter to use as an ECM electrode.
- Zeng et al. (2008) fabricated arrayed microrods of $30\,\mu m$ diameter on an AgW electrode with a $100\,\mu m$ thick Mn foil.
- Mujumdar et al. (2010) fabricated two pillars of $200\,\mu m$ diameter (cylindrical rod) and $400\,\mu m$ side length (rectangular prism) using Cu foils with through holes.
- Singh et al. (2019) fabricated 4×4 arrayed tungsten microrods of $58\,\mu m$ diameter and $830\,\mu m$ length.

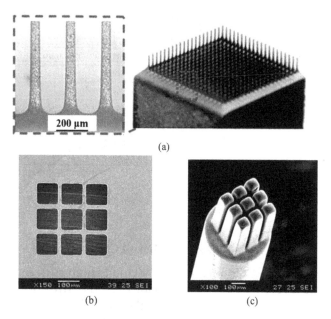

FIGURE 6.13
(a) A 40 × 40 micropin array fabricated using a brass plate of 600 μm thick and 30 μm hole dia (Hwang et al., 2010), (b) AISI 304 foil electrode with square holes, and (c) square rods fabricated with the foil electrode (Yi et al., 2008).

- Roy et al. (2018) machined a 3D hemispherical microfeature on brass using R-EDM.
- Mastud et al. (2015) fabricated a textured surface with micropillars on Ti-6Al-4V.

6.5 Hybrid Micro EDM

To enhance the performance characteristics, researchers combined different micromachining strategies with micro EDM. These combinations can be broadly classified into two types: sequential processing and simultaneous processing. In sequential processing, micro EDM is used as the primary machining strategy, and some special techniques are included in the preprocessing (for tool electrode preparation) and post-processing stage (surface quality improvement). In simultaneous processing, various techniques are clubbed with micro EDM for machining assistance. The classification of hybrid micro EDM techniques is shown in Figure 6.14.

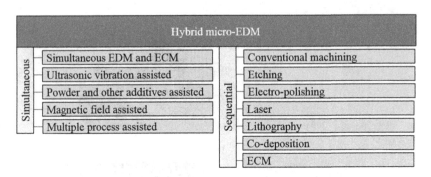

FIGURE 6.14
Overview of hybrid EDM processes.

6.5.1 Simultaneous Micro EDM and Micro ECM

The surface machined with micro ECM exhibits high surface quality due to the atomic-level material removal. However, the MRR of ECM is very low. Micro EDM gives comparatively higher machining rate, but the surface quality is affected by the presence of overlapping recast layers and thermal cracks. Sequentially clubbing these two processes to utilize the advantageous machining characteristics of each process is attempted by various researchers. Establishing a simultaneous ECM-EDM process is challenging due to the unavailability of a single working fluid for facilitating these two processes. Nguyen et al. (2012) presented an innovative way to combine these two machining strategies by selecting low-resistivity deionized water as the working fluid. Deionized water with a resistivity of 0.1–0.5 Ω cm is considered to be partially deionized. A small number of ions that exist in the fluid induce slight conductivity. Deionized water with low resistivity can also be a weak current carrier during the electrochemical reaction, which makes low-resistivity deionized water a perfect working fluid for combining EDM and ECM, as it can induce sparks at higher voltages and facilitate weak electrochemical action. After dielectric breakdown, sparks remove material from the workpiece by melting and vaporization, leaving small craters on the surface. Weak electrochemical dissolution smoothens the peaks created by the overlapping craters, and the surface roughness reduces as shown in Figure 6.15. As the dissolution process is slower than the material removal by discharges, the tool feed has to be adjusted accordingly to incorporate the difference in MRR. The process continues as the tool descends to the workpiece. Figure 6.16 shows a microhole machined with the hybrid ECM-EDM machining technique with a power supply of pulse frequency 500 kHz and 30% duty cycle. The results show a smooth sidewall and pitting-free rim surface (Nguyen et al., 2012).

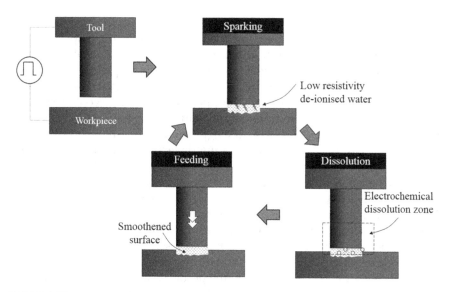

FIGURE 6.15
Simultaneous micro EDM and micro ECM.

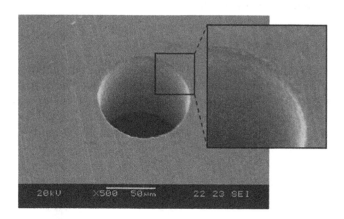

FIGURE 6.16
Microhole fabricated by simultaneous micro ECM and micro EDM (Nguyen et al., 2012).

6.5.2 Vibration-Assisted Micro EDM

The inefficiency of the dielectric flushing system to remove debris particles from the machining zone during EDM often results in dimensional inaccuracies. Vibration-assisted micro EDM is reported as a successful solution towards enhancing debris evacuation from the machining zone. During micro EDM, the debris is often removed by the gaseous bubbles escaping from the machining zone (Yu et al., 2009). As the depth of the hole increases, the viscous resistance increases and the bubbles begin to occupy the hole bottom, halting debris removal and machining. Introduction of ultrasonic vibrations to the machining zone results in high-frequency alternating pressure variation in the sparking zone. This pressure variation helps to remove the molten material from the microcavity. The viscous resistance in the machining zone is reduced significantly, and the bubbles escape out carrying the debris particles. Also, the agglomeration of the debris particles is prevented with the help of the vibrating dielectric fluid. In brief, enhancing the dielectric flushing efficiency and creating pressure variations in the machining zone avoids arcing, secondary discharges, and improves the machining efficiency (Khatri et al., 2016). Ultrasonic vibrations can be introduced to the machining zone by different methods. Yu et al. (2009) used a vibrating tool electrode to machine high-aspect-ratio microholes. Jahan et al. (2011) introduced vibrations to the workpiece. Oschätzchen et al. (2013) used a sonotrode to vibrate the dielectric fluid during micro EDM. All three modes are shown in Figure 6.17. Liew et al. (2014) proposed a method to induce cavitation in the machining zone with the help of a probe-type vibrator. A cloud of cavitation bubbles generated by ultrasonic vibrations is used to carry away the debris particles. Major research works and findings in vibration-assisted micro EDM are listed in Table 6.1.

6.5.3 Magnetic Field-Assisted Micro EDM

Some researchers suggested that EDM under magnetic field can help to evacuate debris particles from the machining zone. The resultant of the centrifugal force and the magnetic force helps to push the debris particles out of the machining zone. Using this principle, Yeo et al. (2004) machined microholes with magnetic field-assisted micro EDM and reported a

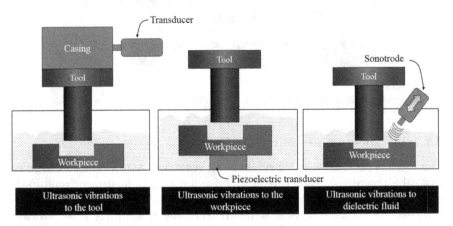

FIGURE 6.17
Methods of introducing ultrasonic vibrations to the machining zone during micro EDM.

TABLE 6.1

Major Research Works in Vibration-Assisted Micro EDM

Authors	Research Findings
Wansheng et al. (2002)	Ultrasonic vibrations help to improve the machining rate and reduce the taper of the hole
Gao and Liu (2003)	Vibrating workpiece improved the machining efficiency by eight times compared to the normal micro EDM
Huang et al. (2003)	Machining Nitinol with the assistance of ultrasonic vibrations helped to improve the machining efficiency by 60 times. The electrode wear is reduced due to increased flushing efficiency
Chern et al. (2006)	Microholes of diameter 100 μm are machined with vibration-assisted micro EDM drilling, and the roundness of the holes is improved
Endo et al. (2012)	A square shaft with a cross section of 50 μm×50 μm is machined using vibration-assisted micro EDM. A notable reduction in machining time, tool jump motions, and increase in normal discharge pulse frequency are reported. Surface roughness and tool wear are not much affected by vibrations
Kim et al. (2006b)	Using variable capacitance and ultrasonic vibrations, a high-aspect-ratio straight hole is machined. Tapering was reduced, but the overcut was near to 7 μm
Hung et al. (2006a)	Using a helical microtool electrode with ultrasonic vibration assistance is useful in reducing the spark gap, taper, and machining time in deep-hole drilling
Sundaram et al. (2008)	Ultrasonic vibrations at 60% peak power are found to be optimum for best machining performance
Tong et al. (2008)	The vibration frequency of 6 kHz with an amplitude of 3 μm is used to machine noncircular microfeatures. The frequency of optimum discharges increases with vibrational frequency
Zeng et al. (2008)	5 × 5 arrays of microelectrodes with size less than 30 μm and aspect ratio 8 are fabricated by ultrasonic vibration-assisted micro EDM. Reduction in total machining time is observed
Endo et al. (2008)	As the frequency and amplitude of vibrations increases, the machining time reduces substantially. Vibration perpendicular to the tool feed direction is found to be more effective than the parallel vibration during EDM. The stiffness of the workpiece is found to be an important parameter during micro EDM with vibration assistance
Hao et al. (2008)	Scanning micro EDM performance is improved with the help of high-frequency vibrations. The MRR reaches up to 1.4×10^5 μm³/s with vibrations of 5 kHz and an amplitude of 2.7 μm
Yu et al. (2009)	A high-aspect-ratio (~29) microhole is drilled with vibration-assisted planetary micro EDM
Jahan et al. (2010a, b, 2011)	Periodic suction and pumping action in the machining zone due to vibrations helps to improve the debris flushing efficiency. An increase in MRR and decrease in tool wear is observed with workpiece vibration. Microholes of aspect ratio up to 16.7 are machined with vibration-assisted (amplitude 1.5 μm and frequency 750 Hz) micro EDM. However, very high frequency and high amplitude vibrations deteriorate the surface quality

(Continued)

TABLE 6.1 (*Continued*)

Major Research Works in Vibration-Assisted Micro EDM

Authors	Research Findings
Garn et al. (2011)	Machining time and the frequency of occurrence of arc discharges decrease with an increase in vibration frequency when the amplitude is kept constant
Mastud et al. (2012a)	Low-frequency vibrations help to increase the percentage of normal discharges in reverse micro EDM. Twenty eight percent reduction in machining time is observed when the amplitude of vibration of the plate electrode is increased from 0.5 to 2 μm
Ichikawa and Natsu (2013)	Compared to normal EDM, the average feed rate in vibration-assisted micro EDM is found to be improved by 33 times. Ultrasonic vibration amplitude has little effect on the machining characteristics
Jafferson et al. (2014)	Ultrasonic vibration improved the machining performance during micro EDM milling as the dielectric fluid velocity increased up to 0.5 m/s. Applying a magnetic field in a vibration-assisted micro EDM system is detrimental to the surface quality
Lee et al. (2015)	Low-frequency vibrations (60 Hz) during micro EDM reduce the machining time by 60% during single-hole machining. For multiple-hole machining, the machining time was reduced by 43% for Cu and 80% for steel. However, vibrations above 60 Hz are having little effect on the machining time
Bajpai et al. (2016)	MRR is increased by 16%, and dimensional errors are decreased by 22% when vibrations of frequency 3 kHz introduced during micro EDM. However, the surface quality is diminished with an increase in vibrational frequency
Li et al. (2016)	Vibration-assisted micro EDM successfully produced clean textured surfaces without surface burns, microcracks, and melted materials. The shape accuracy increased by vibration in the machining zone
Beigmoradi et al. (2018)	Vibration-assisted micro EDM is analyzed by simulating the velocity contour of dielectric fluid for different vibrational frequencies and electrode materials. Tungsten electrode is observed to be having more machining stability due to less lateral movements compared to copper and graphite tool electrodes

26% increase in the achievable depth during high-aspect-ratio drilling. However, the presence of a magnetic field caused deflections to the tool electrode and eventually resulted in shape distortions and dimensional errors. Heinz et al. (2011) carried out a more elaborate study and examined the effect of magnetic field during micro EDM of nonmagnetic materials. Directional workpiece current is used to induce Lorentz force, which acts normal to the workpiece surface. Even though the workpiece debris particles are nonmagnetic, the induced force will push it out from the machining zone and enhance the machining performance as shown in Figure 6.18. The erosion efficiency is increased to 3.5% when a 0.33 T magnetic flux is used, and the efficiency increased to 5.4% when the magnetic flux strength is increased to 0.66 T.

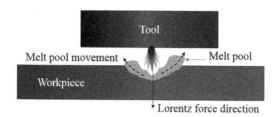

FIGURE 6.18
Principle of using Lorentz force to push away molten pool from the machined crater.

6.5.4 Powder-Assisted Micro EDM

Studies revealed that the addition of various powder particles to the dielectric fluid makes positive effects in the machining performance. This phenomenon is attributed to the increase in electrical and thermal conductivity of the fluid medium, which affects the breakdown characteristics of the fluid. Increase in the spark gap due to a reduction in insulating strength of the dielectric facilitates more efficient dielectric flushing and debris removal. This increases the machining stability during micro EDM. In the spark gap, the particles form a chain-like structure in the direction of current flow. As the dielectric breakdown occurs, part of the discharge energy is uniformly distributed among the particles, and the discharge intensity reduces. This will result in smaller craters and higher surface finish. The expansion of plasma results in reduced plasma pressure. The reduction in pressure causes the retention of the molten pool in the crater. The increased thermal conductivity of the fluid increases the solidification rate of molten material. In brief, the crater becomes smaller with consistent depth, but a material buildup exists at the center and outer rim of the crater. Table 6.2 consolidates some important research works in powder-mixed micro EDM.

TABLE 6.2

Major Research Works in Micro EDM with Powder-Mixed Dielectric Fluids

Authors	Powder Material	Findings
Yeo et al. (2007)	SiC	Craters with shallow and constant depth are produced during machining with powder-added dielectric fluid
Chow et al. (2008)	SiC	Distribution of discharges resulted in more uniform surface. The presence of powder particles resulted in larger slit width and tool wear
Kibria et al. (2010)	B_4C	Adding B_4C to kerosene does not make much difference in MRR. B_4C-mixed deionized water performed well with high MRR due to efficient distribution of discharge. Recast layer thickness is also reduced

(Continued)

TABLE 6.2 (Continued)

Major Research Works in Micro EDM with Powder-Mixed Dielectric Fluids

Authors	Powder Material	Findings
Prihandana et al. (2011)	Nanographite powder	Machining time is reduced by 35% in the presence of graphite nanoparticles because of the stable machining process and increased spark gap. Microcracks on the surface are eliminated with the addition of nanoparticles due to reduced discharge power density
Prihandana et al. (2009a, b, 2014)	MoS_2 powder	Maximum MRR achieved at a powder concentration level of 10 g/l. Fifty nanometers size particles give the highest MRR compared to particles with 10 nm and 2 μm
Tan and Yeo (2011)	SiC	A 15%–35% reduction in recast layer is observed while machining with Idemitsu Daphne™ Cut HL-25 dielectric fluid with 0.1 g/l of powder concentration
Cyril et al. (2017)	Al, Gr, SiC	Tool wear decreased during machining due to enhanced heat dissipation in the presence of powder particles. Material migration observed on the workpiece surface due to the implosion of gases in the interelectrode gap
Elsiti and Noordin (2017)	Fe_2O_3	MRR increased when the powder concentration is increased. But at higher concentration, MRR reduced due to powder settling problem and bridging effects
Modica et al. (2018)	Garnet	Deionized and tap water mixed with Garnet showed deposited layers on the workpiece
Tiwary et al. (2018)	Cu powder	At higher peak current, powder-mixed deionized water showed minimum overcut and improved circularity. The white layer is not observed during machining with Cu powder-mixed dielectric fluid

6.5.5 Sequential Micro EDM and Laser Machining

Combining laser micromachining and micro EDM in a sequential strategy helps to increase MRR and surface quality. Laser micromachining is known for its high machining rate and poor surface quality. On the other hand, micro EDM is a slower process with high machining accuracy and capable of producing high-quality machined surfaces. In the sequential method, the microfeature is machined with laser machining first, and micro EDM is used to improve the dimensional accuracy and surface quality as shown in Figure 6.19. Al-Ahmari et al. (2016) machined microholes on Ni-Ti-based shape memory alloy using sequential laser and EDM. The comparison of the microholes drilled by different methods is shown in Figure 6.20. The taper and recast layer formation are reduced by combined machining. The MRR is improved by 40%–65% during hybrid machining (Al-Ahmari et al., 2016). Premachining of microfeatures using a laser during micro EDM milling and drilling is studied by Kim et al. (2010). The drilling time is reduced by 90%, and the milling time is reduced by 75% with the usage of laser premachining.

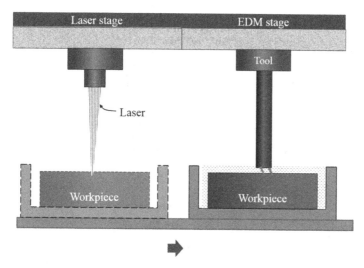

FIGURE 6.19
Sequential micro EDM and laser machining.

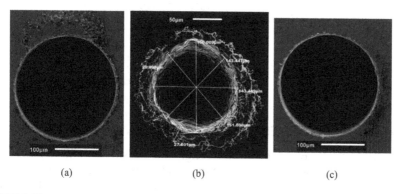

(a)   (b)   (c)

FIGURE 6.20
Microhole machined with (a) normal micro EDM, (b) laser micromachining, and (c) hybrid method (Al-Ahmari et al., 2016).

6.5.6 Sequential Micro EDM and LIGA

In a hybrid method combining LIGA and micro EDM, a series of microholes of different geometries can be fabricated as shown in Figure 6.21 (Takahata et al., 2000; Hu et al., 2009). Micro EDM can be used for postprocessing the microcomponents fabricated using soft lithography technique (Essa et al., 2017). Surface roughness and flatness of the microparts fabricated by lithography are improved by subsequent surface processing with low-energy electric discharges.

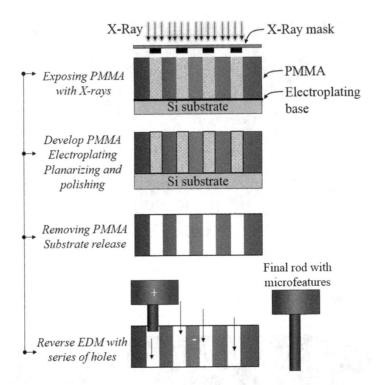

FIGURE 6.21
Sequential hybrid machining process combining LIGA and micro EDM.

6.5.7 Sequential Micro EDM, Electroforming, and Etching

Combining micro EDM, electroforming, and selective etching processes, microfeatures can be fabricated as shown in Figure 6.22 (Tosello et al., 2008). Micro EDM and electroforming are used to fabricate a microball joint for the application in microrobotic arms (Lin et al., 2010). Combination of micro EDM and electrochemical etching can be used to fabricate indium tin oxide line patterns on a glass substrate (Trinh et al., 2017).

6.5.8 Other Hybrid Processes

- Microturning is used as a tool fabrication technique during micro EDM drilling (Asad et al., 2007).
- Electropolishing is used as a postprocessing operation to improve the surface quality of microholes drilled with micro EDM (Hung et al., 2006b).

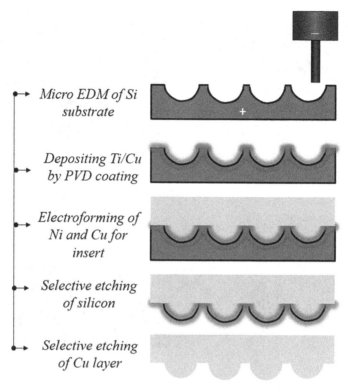

Micro EDM of Si
substrate

Depositing Ti/Cu
by PVD coating

Electroforming of
Ni and Cu for
insert

Selective etching
of silicon

Selective etching
of Cu layer

FIGURE 6.22
Steps in hybrid machining process are combining micro EDM, electroforming, and etching.

- Micro ECM is used as sequential operation after micro EDM milling to improve the surface finish of the microfeatures (Chung et al., 2014; He et al., 2013; Zeng et al., 2012).
- Micro EDM and codeposition method are used to fabricate microtools for simultaneous drilling and grinding (Chen et al., 2008; Hsue and Chang, 2016; Hung et al., 2007).

6.5.9 Applications of Hybrid Micro EDM Processes

- Sequential micro EDM followed by electrochemical finishing is used to fabricate microfeatures (Figure 6.23a) where the surface roughness (R_a) is reduced from 0.707 to 0.143 µm (Zeng et al., 2012).
- Combination of micro EDM and laser is used for a microassembly where micro EDM is used for microfeature fabrication, and the laser is used for microwelding (Huang and Kuo, 2002; Liang et al., 2002).

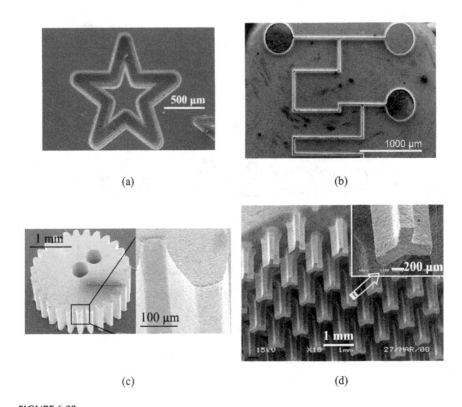

FIGURE 6.23
Microcomponents fabricated by combining micro EDM with (a) micro ECM (Zeng et al., 2012), (b) laser machining (Shiu et al., 2010), (c) soft lithography (Essa et al., 2017), and (d) UV-LIGA process (Hu et al., 2009).

- Hybrid micro EDM–laser machining is used to fabricate the fuel injection nozzle, where the drilling time is reduced by 70% and the total cost is reduced by 42% (Li et al., 2006).
- Metallic micromolds for fabricating polymeric microfluidic devices (Figure 6.23b) are machined with micro EDM–laser hybrid machining (Shiu et al., 2010).
- Microgears with minimum edge distortions (Figure 6.23c) are machined with sequential operation of soft lithography and micro EDM (Essa et al., 2017).
- Complex-shaped microrods (Figure 6.23d) are fabricated by combining ultraviolet (UV) LIGA with micro EDM (Hu et al., 2009).
- Microgas turbine motor rotor is fabricated by micro EDM and extrusion (Geng et al., 2014).

6.6 Other Special Variants of Micro EDM

6.6.1 Micro EDM Deposition

Erosion of the tool electrode is an inevitable phenomenon during micro EDM. Some researchers utilized tool erosion microfeature fabrication on workpiece surface using material deposition. Here, the tool electrode is connected as the anode, and a low discharge energy power supply for a short duration is applied between the electrodes. Air is used as the dielectric medium. As the process continues, the tool electrode material is melted and deposited on the workpiece surface. By controlling axis movements, researchers fabricated different microstructures using micro EDM deposition (Ori et al., 2004). Wang et al. (2009) fabricated an array of tungsten microelectrodes and a brass microspiral structure. Jin et al. (2007) fabricated steel and brass microcylinders. A 22-circle spiral structure with a diameter of 190 μm and a height of 3.39 mm is also fabricated with the deposition method (Chi et al., 2008).

6.6.2 Deburring and Surface Modifications

Removing microburrs formed during conventional micromilling or drilling operation is very challenging. Micro EDM is capable of removing burrs without causing defections or dimensional errors to the microfeatures (Jeong et al., 2009). Micro EDM is also used for deburring of microholes drilled (by conventional drilling) on carbon fiber reinforced plastic (CFRP) composites (Islam et al., 2017). Beyond deburring action, micro EDM can also be used for nanofinishing the workpiece surfaces (Jahan et al., 2009). The surface roughness can be brought under $R_a = 0.06$ μm by micro EDM with a special orbiting strategy (Maradia et al., 2013). Micro EDM can also be used for surface modifications, as the machined surface exhibits higher hardness than the parent workpiece surface due to recast layer formation and carbon deposition (Prakash et al., 2017).

6.6.3 Nanoparticle Fabrication

The nanometer-sized debris particles are flushed out from the machining zone during micro EDM. This phenomenon can be utilized to fabricate nanoparticles of different materials. A workpiece with a particular material composition is placed as an anode during micro EDM. After machining, the debris particles are collected from the tank bottom. A work tank with a sonicator avoids the chance of coagulation of debris particles. Sahu et al. (2014) fabricated copper nanoparticles using micro EDM. Tseng et al. (2016) developed a new micro EDM system dedicated to the fabrication of nanosilver colloid. The same principle is applied for fabrication of nickel nanoparticles with an average crystal size of 15–20 nm (Kumar et al., 2017).

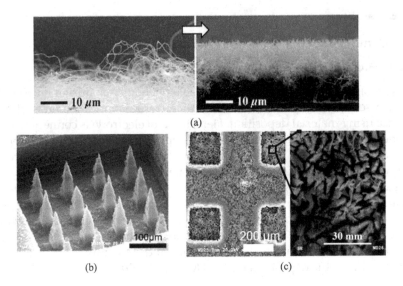

FIGURE 6.24
(a) Enhancing the uniformity of CNT forests (Ok et al., 2007), (b) microstructuring of CNC forests with micro EDM (Dahmardeh et al., 2011b), and (c) rectangular microfeatures fabricated on CNT forests using LIGA-assisted micro EDM (Sarwar et al., 2014).

6.6.4 Machining of Carbon Nanotube Forests

Micro EDM is used for enhancing the field emission uniformity (Kim et al., 2007; Ok et al., 2007) and microstructuring (Dahmardeh et al., 2011b) of carbon nanotube (CNT) forests. Air or gas is used as the dielectric medium during machining to avoid the shrinkage of forests and destruction of the microstructure after drying. This method is capable of fabricating 3D microstructures on the CNT platform, which is very challenging to achieve by other micropatterning methods. Figure 6.24a shows the result of uniformity enhancement with the help of EDM discharges (Ok et al., 2007). Microstructures with an aspect ratio of 20 and a resolution of 5 μm with controlled sidewall angles are fabricated by micro EDM as shown in Figure 6.24b (Dahmardeh et al., 2011a, b). Figure 6.24c shows the rectangular patterns on the CNT forests fabricated via LIGA-assisted micro EDM (Sarwar et al., 2014).

6.7 Conclusion

EDG, R-EDM, planetary EDM, and various hybrid EDM techniques help to expand the capabilities of micro EDM. On-machine tool fabrication can be realized using the EDG process, which reduces tool handling errors. Dielectric flushing is enhanced by planetary EDM. An array of microelectrodes can be fabricated easily using R-EDM method. Combining other machining strategies with EDM helps to exploit the advantages of different machining methods. Vibration-assisted micro EDM increases the percentage of normal discharges. Powder-assisted micro EDM makes a smaller crater and provides a higher surface finish. Combining ECM, laser machining, LIGA, and etching with EDM as a simultaneous or sequential machining operation helps to improve the characteristics of the machined surface.

References

Al-Ahmari, A.M.A., Rasheed, M.S., Mohammed, M.K., Saleh, T., 2016. A hybrid machining process combining micro EDM and laser beam machining of nickel-titanium-based shape memory alloy. *Mater. Manuf. Process.* 31, 447–455.

Ali, M.Y., Ong, A.S., 2006. Fabricating micromilling tool using wire electrodischarge grinding and focused ion beam sputtering. *Int. J. Adv. Manuf. Technol.* 31, 501–508.

Asad, A.B.M.A., Masaki, T., Rahman, M., Lim, H.S., Wong, Y.S., 2007. Tool-based micro-machining. *J. Mater. Process. Technol.* 192–193, 204–211.

Bajpai, V., Mahambare, P., Singh, R.K., 2016. Effect of thermal and material anisotropy of pyrolytic carbon in vibration-assisted micro EDM process. *Mater. Manuf. Process.* 31, 1879–1888.

Bamberg, E., Heamawatanachai, S., 2008. Orbital electrode actuation to improve efficiency of drilling micro-holes by micro EDM. *J. Mater. Proc. Technol.* 9, 1826–1834. doi: 10.1016/j.jmatprotec.2008.04.044.

Beigmoradi, S., Ghoreishi, M., Vahdati, M., 2018. Optimum design of vibratory electrode in micro EDM process. *Int. J. Adv. Manuf. Technol.* 95, 3731–3744.

Chen, S.T., Lai, Y.C., Liu, C.C., 2008. Fabrication of a miniature diamond grinding tool using a hybrid process of micro EDM and co-deposition. *J. Micromech. Microeng.* 18, 055005.

Chern, G.L., de Wang, S., 2007. Punching of noncircular micro-holes and development of micro-forming. *Precis. Eng.* 31, 210–217.

Chern, G.L., Wu, Y.J.E., Cheng, J.C., Yao, J.C., 2007. Study on burr formation in micro-machining using micro-tools fabricated by micro EDM. *Precis. Eng.* 31, 122–129.

Chern, G.L., Wu, Y.J.E., Liu, S.F., 2006. Development of a micro-punching machine and study on the influence of vibration machining in micro EDM. *J. Mater. Process. Technol.* 180, 102–109.

Chi, G., Wang, Z., Xiao, K., Cui, J., Jin, B., 2008. The fabrication of a micro-spiral structure using EDM deposition in the air. *J. Micromech. Microeng.* 18, 1–9.

Chow, H.M., Yang, L.D., Lin, C.T., Chen, Y.F., 2008. The use of SiC powder in water as dielectric for micro-slit EDM machining. *J. Mater. Process. Technol.* 195, 160–170.

Chung, D.K., Lee, K.H., Jeong, J., Chu, C.N., 2014. Machining characteristics on electrochemical finish combined with micro EDM using deionized water. *Int. J. Precis. Eng. Manuf.* 15, 1785–1791.

Cyril, J., Paravasu, A., Jerald, J., Sumit, K., Kanagaraj, G., 2017. Experimental investigation on performance of additive mixed dielectric during micro-electric discharge drilling on 316L stainless steel. *Mater. Manuf. Processes.* 32(6), 638–644.

Dahmardeh, M., Khalid, W., Mohamed Ali, M.S., Choi, Y., Yaghoobi, P., Nojeh, A., Takahata, K., 2011a. High-aspect-ratio, 3-D micromachining of carbon-nanotube forests by micro-electro-discharge machining in air. *Proceedings of the IEEE International Conference on Micro Electro Mechanical Systems*, Cancun, Mexico, 272–275.

Dahmardeh, M., Nojeh, A., Takahata, K., 2011b. Possible mechanism in dry micro-electro-discharge machining of carbon-nanotube forests: A study of the effect of oxygen. *J. Appl. Phys.* 109, 093308.

Egashira, K., Taniguchi, T., Hanajima, S., Tsuchiya, H., Miyazaki, M., 2006. Planetary EDM of micro holes. *Int. J. Electr. Mach.* 11, 15–18.

Elsiti, N.M., Noordin, M.Y., 2017. Experimental investigations into the effect of process parameters and nano-powder on material removal rate during micro EDM of Co-Cr-Mo. *Key Eng. Mater.* 740, 125–132.

Endo, T., Tsujimoto, T., Mitsui, K., 2008. Study of vibration-assisted micro EDM—the effect of vibration on machining time and stability of discharge. *Precis. Eng.* 32, 269–277.

Endo, T., Tsujimoto, T., Mitsui, K., 2012. Fabrication of micro-components using vibration- assisted micro- EDM School of Integrated Design Engineering Department of Mechanical Engineering, 1–4.

Essa, K., Modica, F., Imbaby, M., El-Sayed, M.A., ElShaer, A., Jiang, K., Hassanin, H., 2017. Manufacturing of metallic micro-components using hybrid soft lithography and micro-electrical discharge machining. *Int. J. Adv. Manuf. Technol.* 91, 445–452.

Fleischer, J., Masuzawa, T., Schmidt, J., Knoll, M., 2004. New applications for micro EDM. *J. Mater. Process. Technol.* 149, 246–249.

Gao, C., Liu, Z., 2003. A study of ultrasonically aided micro-electrical-discharge machining by the application of workpiece vibration. *J. Mater. Process. Technol.* 139, 226–228.

Garn, R., Schubert, A., Zeidler, H., 2011. Analysis of the effect of vibrations on the micro EDM process at the workpiece surface. *Precis. Eng.* 35, 364–368.

Geng, X., Chi, G., Wang, Y., Wang, Z., 2014. High-efficiency approach for fabricating MTE rotor by micro EDM and micro-extrusion. *Chin. J. Mech. Eng.* 27, 830–835.

Guo, X., Yu, Z., Lv, Z., Li, J., Natsu, W., 2013. Optimization of planetary movement parameters for microhole drilling by micro-electrical discharge machining. *J. Micro Nano Manuf.* 1, 031007.

Hao, T., Yang, W., Yong, L., 2008. Vibration-assisted servo scanning 3D micro EDM. *J. Micromech. Microeng.* 18, 025011.

He, X.L., Wang, Y.K., Wang, Z.L., Zeng, Z.Q., 2013. Micro-hole drilled by EDM-ECM combined processing. *Key Eng. Mater.* 562–565, 52–56.

Heinz, K., Kapoor, S.G., DeVor, R.E., Surla, V., 2011. An investigation of magnetic-field-assisted material removal in micro EDM for nonmagnetic materials. *J. Manuf. Sci. Eng.* 133, 021002.

Hsue, A.W.J., Chang, Y.F., 2016. Toward synchronous hybrid micro EDM grinding of micro-holes using helical taper tools formed by Ni-Co/diamond Co-deposition. *J. Mater. Process. Technol.* 234, 368–382.

Hu, Y.Y., Zhu, D., Qu, N.S., Zeng, Y.B., Ming, P.M., 2009. Fabrication of high-aspect-ratio electrode array by combining UV-LIGA with micro electro-discharge machining. *Microsyst. Technol.* 15, 519–525.

Huang, H., Zhang, H., Zhou, L., Zheng, H.Y., 2003. Ultrasonic vibration assisted electro-discharge machining of microholes in Nitinol. *J. Micromech. Microeng.* 13, 693–700.

Huang, J.D., Kuo, C.L., 2002. Pin-plate micro assembly by integrating micro EDM and Nd-YAG laser. *Int. J. Mach. Tools Manuf.* 42, 1455–1464.

Hung, J.-C., Wu, W.-C., Yan, B.-H., Huang, F.-Y., Wu, K.-L., 2007. Fabrication of a micro-tool in micro EDM combined with co-deposited Ni–SiC composites for micro-hole machining. *J. Micromech. Microeng.* 17, 763–774.

Hung, J.C., Lin, J.K., Yan, B.H., Liu, H.S., Ho, P.H., 2006a. Using a helical micro-tool in micro EDM combined with ultrasonic vibration for micro-hole machining. *J. Micromech. Microeng.* 16, 2705–2713.

Hung, J.C., Yan, B.H., Liu, H.S., Chow, H.M., 2006b. Micro-hole machining using micro EDM combined with electropolishing. *J. Micromech. Microeng.* 16, 1480–1486.

Hwang, Y.L., Kuo, C.L., Hwang, S.F., 2010. Fabrication of a micro-pin array with high density and high hardness by combining mechanical peck-drilling and reverse-EDM. *J. Mater. Process. Technol.* 210, 1103–1130.

Ichikawa, T., Natsu, W., 2013. Realization of micro EDM under ultra-small discharge energy by applying ultrasonic vibration to machining fluid. *Procedia CIRP 6*, 326–331.

Islam, M.M., Li, C.P., Won, S.J., Ko, T.J., 2017. A deburring strategy in drilled hole of CFRP composites usingï¿½EDMï¿½process. *J. Alloys Compd.* 703, 477–485.

Jafferson, J.M., Hariharan, P., Ram Kumar, J., 2014. Effects of ultrasonic vibration and magnetic field in micro EDM milling of nonmagnetic material. *Mater. Manuf. Process.* 29, 357–363.

Jahan, M.P., Anwar, M.M., Wong, Y.S., Rahman, M., 2009. Nanofinishing of hard materials using micro-electrodischarge machining. *Proc. Inst. Mech. Eng. Part B J. Eng. Manuf.* 223, 1127–1142.

Jahan, M.P., Rahman, M., Wong, Y.S., Fuhua, L., 2010a. On-machine fabrication of high-aspect-ratio micro-electrodes and application in vibration-assisted micro-electrodischarge drilling of tungsten carbide. *Proc. Inst. Mech. Eng. Part B J. Eng. Manuf.* 224, 795–814.

Jahan, M.P., Saleh, T., Rahman, M., Wong, Y.S., 2011. Study of micro EDM of tungsten carbide with workpiece vibration. *Adv. Mater. Res.* 264–265, 1056–1061.

Jahan, M.P., Saleh, T., Rahman, M., Wong, Y.S., 2010b. Development, modeling, and experimental investigation of low frequency workpiece vibration-assisted micro EDM of tungsten carbide. *J. Manuf. Sci. Eng.* 132, 54503.

Jeong, Y.H., HanYoo, B., Lee, H.U., Min, B.K., Cho, D.W., Lee, S.J., 2009. Deburring microfeatures using micro EDM. *J. Mater. Process. Technol.* 209, 5399–5406.

Jin, B.D., Cao, G.H., Wang, Z.L., Zhao, W.S., 2007. A micro-deposition method by using EDM. *Key Eng. Mater.* 339, 32–36.

Khatri, B.C., Rathod, P., Valaki, J.B., 2016. Ultrasonic vibration-assisted electric discharge machining: A research review. *Proc. Inst. Mech. Eng. Part B J. Eng. Manuf.* 230, 319–330.

Kibria, G., Sarkar, B.R., Pradhan, B.B., Bhattacharyya, B., 2010. Comparative study of different dielectrics for micro EDM performance during microhole machining of Ti-6Al-4V alloy. *Int. J. Adv. Manuf. Technol.* 48, 557–570.

Kim, B.H., Park, B.J., Chu, C.N., 2006a. Fabrication of multiple electrodes by reverse EDM and their application in micro ECM. *J. Micromech. Microeng.* 16, 843.

Kim, D.J., Yi, S.M., Lee, Y.S., Chu, C.N., 2006b. Straight hole micro EDM with a cylindrical tool using a variable capacitance method accompanied by ultrasonic vibration. *J. Micromech. Microeng.* 16, 1092–1097.

Kim, G.B.H., Ok, J.G., Kim, Y.H., Chu, C.N., 2007. Electrical discharge machining of carbon nanofiberfor uniform field emission. *CIRP Ann. Manuf. Technol.* 56, 233–236.

Kim, S., Kim, B.H., Chung, D.K., Shin, H.S., Chu, C.N., 2010. Hybrid micromachining using a nanosecond pulsed laser and micro EDM. *J. Micromech. Microeng.* 20, S276–S280.

Kumar, P., Singh, P.K., Kumar, D., Prakash, V., Hussain, M., Das, A.K., 2017. A novel application of micro EDM process for the generation of nickel nanoparticles with different shapes. *Mater. Manuf. Process.* 32, 564–572.

Lee, P.A., Kim, Y., Kim, B.H., 2015. Effect of low frequency vibration on micro EDM drilling. *Int. J. Precis. Eng. Manuf.* 16, 2617–2622.

Li, L., Diver, C., Atkinson, J., Giedl-Wagner, R., Helml, H.J., 2006. Sequential laser and EDM micro-drilling for next generation fuel injection nozzle manufacture. *CIRP Ann. Manuf. Technol.* 55, 179–182.

Li, Y., Deng, J., Chai, Y., Fan, W., 2016. Surface textures on cemented carbide cutting tools by micro EDM assisted with high-frequency vibration. *Int. J. Adv. Manuf. Technol.* 82, 2157–2165.

Liang, H.Y., Kuo, C.L., Huang, J.D., 2002. Precise micro-assembly through an integration of micro EDM and Nd-YAG. *Int. J. Adv. Manuf. Technol.* 20, 454–458.

Liew, P.J., Yan, J., Kuriyagawa, T., 2014. Fabrication of deep micro-holes in reaction-bonded SiC by ultrasonic cavitation assisted micro EDM. *Int. J. Mach. Tools Manuf.* 76, 13–20.

Lin, C.S., Liao, Y.S., Cheng, Y.T., Lai, Y.C., 2010. Fabrication of micro ball joint by using micro EDM and electroforming. *Microelectron. Eng.* 87, 1475–1478.

Maradia, U., Scuderi, M., Knaak, R., Boccadoro, M., Beltrami, I., Stirnimann, J., Wegener, K., 2013. Super-finished surfaces using meso-micro EDM. *Procedia CIRP* 6, 157–162.

Masaki, T., Kuriyagawa, T., Yan, J., Yoshihara, N., 2007. Study on shaping spherical poly crystalline diamond tool by micro-electro-discharge machining and micro-grinding with the tool. *Int. J. Surf. Sci. Eng.* 1, 344.

Mastud, S., Garg, M., Singh, R., Samuel, J., Joshi, S., 2012a.Experimental characterization of vibration-assisted reverse micro electrical discharge machining (EDM) for surface texturing. *ASME 2012 International Manufacturing Science and Engineering Conference (MSEC 2012)*, Indiana, 1–10.

Mastud, S., Singh, R.K., Joshi, S.S., 2012b. Analysis of fabrication of arrayed micro-rods on tungsten carbide using reverse micro EDM. *Int. J. Manuf. Technol. Manag.* 26, 176–195.

Mastud, S., Singh, R.K., Samuel, J., Joshi, S.S., 2011. Comparative analysis of the process mechanics in micro-electrical discharge machining (EDM) and reverse micro EDM. *ASME 2011 International Manufacturing Science and Engineering Conference.* American Society of Mechanical Engineers, Oregon, 349–358.

Mastud, S.A., Garg, M., Singh, R., Joshi, S.S., 2012c. Recent developments in the reverse micro-electrical discharge machining in the fabrication of arrayed micro-features. *Proc. Inst. Mech. Eng. Part C J. Mech. Eng. Sci.* 226, 367–384.

Mastud, S.A., Kothari, N.S., Singh, R.K., Joshi, S.S., 2015. Modeling debris motion in vibration assisted reverse micro electrical discharge machining process (R-MEDM). *J. Microelectromech. Syst.* 24, 661–676.

Masuzawa, T., Fujino, M., Kobayashi, K., Suzuki, T., Kinoshita, N., 1985. Wire electro-discharge grinding for micro-machining. *CIRP Ann. Manuf. Technol.* 34, 431–434.

Modica, F., Marrocco, V., Valori, M., Viganò, F., Annoni, M., Fassi, I., 2018. Study about the influence of powder mixed water based fluid on micro EDM process. *Procedia CIRP* 68, 789–795.

Morgan, C.J., Vallance, R.R., Marsh, E.R., 2004. Micro machining glass with poly crystalline diamond tools shaped by micro electro discharge machining. *J. Micromech. Microeng.* 14, 1687–1692.

Morgan, C.J., Vallance, R.R., Marsh, E.R., 2006. Micro-machining and micro-grinding with tools fabricated by micro electro-discharge machining. *Int. J. Nanomanuf.* 1, 242–258.

Mujumdar, S.S., Mastud, S.A., Singh, R.K., Joshi, S.S., 2010. Experimental characterization of the reverse micro-electrodischarge machining process for fabrication of high-aspect-ratio micro-rod arrays. *Proc. Inst. Mech. Eng. Part B J. Eng. Manuf.* 224, 777–794.

Nguyen, M.D., Rahman, M., Wong, Y.S., 2012. Simultaneous micro EDM and micro-ECM in low-resistivity deionized water. *Int. J. Mach. Tools Manuf.* 54–55, 55–65.

Ok, J.G., Kim, B.H., Sung, W.Y., Chu, C.N., Kim, Y.H., 2007. Uniformity enhancement of carbon nanofiber emitters via electrical discharge machining. *Appl. Phys. Lett.* 90, 1–4.

Oliaei, S.N.B., Özdemir, C., Karpat, Y., 2013. Fabrication of micro ball end mills using micro electro discharge machining. *7th International Conference and Exhibition on Design and Production of Machines and Dies*, Antalya, Turkey.

Ori, R.I., Itoigawa, F., Hayakawa, S., Nakamura, T., Tanaka, S.-I., 2004. Development of advanced alloying process using micro EDM deposition process. *ASME 7th Biennial Conference on Engineering Systems Design and Analysis*, Manchester, UK. The American Society of Mechanical Engineers, 365–370.

Oschätzchen, M.H., Schubert, A., Schneider, J., Zeidler, H., Hahn, M., 2013. Enhancing micro EDM using ultrasonic vibration and approaches for machining of non-conducting ceramics. *Strojniški Vestn. J. Mech. Eng.* 59, 156–164.

Penache, C., Gessner, C., Brauning-Demian, A., Scheffler, P., Spielberger, L., Hohn, O., Schossler, S., Jahnke, T., Gericke, K.-H., Schmidt-Boecking, H.W., 2002. Microstructured electrode arrays: A source of high-pressure nonthermal plasma, in: V.N. Ochkin, ed. *Selected Research Papers on Spectroscopy of Nonequilibrium Plasma at Elevated Pressures*. International Society for Optics and Photonics, Bellingham, WA, pp. 17–26.

Perveen, A., Jahan, M.P., Rahman, M., Wong, Y.S., 2012. A study on microgrinding of brittle and difficult-to-cut glasses using on-machine fabricated poly crystalline diamond (PCD) tool. *J. Mater. Process. Technol.* 212, 580–593.

Prakash, V., Shubham, Singh, P.K., Das, A.K., Chattopadhyaya, S., Mandal, A., Dixit, A.R., 2017. Surface alloying of miniature components by micro-electrical discharge process. *Mater. Manuf. Process.* 33, 1–11.

Prihandana, G.S., Mahardika, M., Hamdi, M., Mitsui, K., 2009a. Effect of m micro MoS$_2$ powder mixed dielectric fluid on surface quality and material removal rate in micro EDM-processes. *Trans. Mater. Res. Soc. Jpn.* 34, 329–332.

Prihandana, G.S., Mahardika, M., Hamdi, M., Wong, Y.S., Mitsui, K., 2011. Accuracy improvement in nanographite powder-suspended dielectric fluid for micro-electrical discharge machining processes. *Int. J. Adv. Manuf. Technol.* 56, 143–149.

Prihandana, G.S., Mahardika, M., Hamdi, M., Wong, Y.S., Mitsui, K., 2009b. Effect of micro-powder suspension and ultrasonic vibration of dielectric fluid in micro EDM processes-Taguchi approach. *Int. J. Mach. Tools Manuf.* 49, 1035–1041.

Prihandana, G.S., Sriani, T., Mahardika, M., Hamdi, M., Miki, N., Wong, Y.S., Mitsui, K., 2014. Application of powder suspended in dielectric fluid for fine finish micro EDM of Inconel 718. *Int. J. Adv. Manuf. Technol.* 75, 599–613.

Qingfeng, Y., Xingqiao, W., Ping, W., Zhiqiang, Q., Lin, Z., Yongbin, Z., 2016. Fabrication of micro rod electrode by electrical discharge grinding using two block electrodes. *J. Mater. Process. Technol.* 234, 143–149.

Rahman, M., Lim, H.S., Neo, K.S., Senthil Kumar, A., Wong, Y.S., Li, X.P., 2007. Tool-based nanofinishing and micromachining. *J. Mater. Process. Technol.* 185, 2–16.

Ravi, N., Huang, H., 2002. Fabrication of symmetrical section microfeatures using the electro-discharge machining block electrode method. *J. Micromech. Microeng.* 12, 905–910.

Roy, T., Datta, D., Balasubramaniam, R., 2018. Reverse micro EDMed 3D hemispherical protruded micro feature: Microstructural and mechanical characterization. *Mater. Res. Express* 6(3), 036513.

Sahu, R.K., Hiremath, S.S., Manivannan, P. V., Singaperumal, M., 2014. Generation and characterization of copper nanoparticles using micro-electrical discharge machining. *Mater. Manuf. Process.* 29, 477–486.

Sarwar, M.S.U., Dahmardeh, M., Nojeh, A., Takahata, K., 2014. Batch-mode micropatterning of carbon nanotube forests using UV-LIGA assisted micro-electro-discharge machining. *J. Mater. Process. Technol.* 214, 2537–2544.

Shiu, P.P., Knopf, G.K., Ostojic, M., 2010. Fabrication of metallic micromolds by laser and electro-discharge micromachining. *Microsyst. Technol.* 16, 477–485.

Singh, A.K., Patowari, P.K., Deshpande, N. V., 2019. Analysis of micro-rods machined using reverse micro EDM. *J. Braz. Soc. Mech. Sci. Eng.* 41, 1–12. doi: 10.1007/s40430-018-1519-4.

Sundaram, M.M., Pavalarajan, G.B., Rajurkar, K.P., 2008. A study on process parameters of ultrasonic assisted micro EDM based on Taguchi method. *J. Mater. Eng. Perform.* 17, 210–215.

Takahata, K., Shibaike, N., Guckel, H., 2000. High-aspect-ratio WC-Co microstructure produced by the combination of LIGA and micro EDM. *Microsyst. Technol.* 6, 175–178.

Tan, P.C., Yeo, S.H., 2011. Investigation of recast layers generated by a powder-mixed dielectric micro electrical discharge machining processg. *Proc. Inst. Mech. Eng. Part B J. Eng. Manuf.* 225, 1051–1062.

Tiwary, A.P., Pradhan, B.B., Bhattacharyya, B., 2018. Investigation on the effect of dielectrics during micro-electro-discharge machining of Ti-6Al-4V. *Int. J. Adv. Manuf. Technol.* 95, 861–874.

Tong, H., Li, Y., Wang, Y., 2008. Experimental research on vibration assisted EDM of micro-structures with non-circular cross-section. *J. Mater. Process. Technol.* 208, 289–298.

Tosello, G., Bissacco, G., Tang, P.T., Hansen, H.N., Nielsen, P.C., 2008. High aspect ratio micro tool manufacturing for polymer replication using μeDM of silicon, selective etching and electroforming. *Microsyst. Technol.* 14, 1757–1764.

Trinh, X.L., Duong, T.H., Kim, H.C., 2017. Large area controllable ITO patterning using micro EDM and electrochemical etching. *Int. J. Adv. Manuf. Technol.* 89, 3681–3689.

Tseng, K.H., Kao, Y.S., Chang, C.Y., 2016. Development and implementation of a micro-electric discharge machine: Real-time monitoring system of fabrication of nanosilver colloid. *J. Clust. Sci.* 27, 763–773.

Wang, M.H., Zhu, D., 2009. Fabrication of multiple electrodes and their application for micro-holes array in ECM. *Int. J. Adv. Manuf. Technol.* 41, 42–47. doi: 10.1007/s00170-008-1456-y.

Wang, Z.L., Cao, G.H., Zhao, W.S., Xiao, K., Jin, B.D., 2009. Fabrication of micro structure using EDM deposition. *Mater. Sci. Forum* 532–533, 305–308.

Wansheng, Z., Zhenlong, W., Shichun, D., Guanxin, C., Hongyu, W., 2002. Ultrasonic and electric discharge machining to deep and small hole on titanium alloy. *J. Mater. Process. Technol.* 120, 101–106.

Weiliang, Z., Zhenlong, W., Desheng, D., 2006. A new micro EDM reverse copying technology for microelectrode array fabrication. *Proceedings of the International Technology and Innovation Conference 2006 (ITIC 2006)*, London, UK.

Xu, J., Guo, B., Shan, D., Wang, Z., Li, M., Fei, X., 2014. Micro-punching process of stainless steel foil with micro-die fabricated by micro EDM. *Microsyst. Technol.* 20, 83–89.

Yeo, S.H., Murali, M., Cheah, H.T., 2004. Magnetic field assisted micro electro-discharge machining. *J. Micromech. Microeng.* 14, 1526–1529.

Yeo, S.H., Tan, P.C., Kurnia, W., 2007. Effects of powder additives suspended in dielectric on crater characteristics for micro electrical discharge machining. *J. Micromech. Microeng.* 17, N91–N98.

Yi, S.M., Park, M.S., Lee, Y.S., Chu, C.N., 2008. Fabrication of a stainless steel shadow mask using batch mode micro EDM. *Microsyst. Technol.* 14, 411–417. doi: 10.1007/s00542-007-0468-0.

Yu, Z.Y., Rajurkar, K.P., Shen, H., 2002. High aspect ratio and complex shaped blind micro holes by micro EDM. *CIRP Ann. Manuf. Technol.* 51, 359–362.

Yu, Z.Y., Zhang, Y., Li, J., Luan, J., Zhao, F., Guo, D., 2009. High aspect ratio micro-hole drilling aided with ultrasonic vibration and planetary movement of electrode by micro EDM. *CIRP Ann. Manuf. Technol.* 58, 213–216. doi: 10.1016/j.cirp.2009.03.111.

Zeng, W.L., Gong, Y.P., Liu, Y., Wang, Z.L., 2008. Experimental study of microelectrode array and micro-hole array fabricated by ultrasonic enhanced micro EDM. *Key Eng. Mater.* 364, 482–487.

Zeng, Z., Wang, Y., Wang, Z., Shan, D., He, X., 2012. A study of micro EDM and micro-ECM combined milling for 3D metallic micro-structures. *Precis. Eng.* 36, 500–509.

Zhan, Z., He, N., Li, L., Shrestha, R., Liu, J., Wang, S., 2015. Precision milling of tungsten carbide with micro PCD milling tool. *Int. J. Adv. Manuf. Technol.* 77, 2095–2103.

Zhang, Z., Peng, H., Yan, J., 2013. Micro-cutting characteristics of EDM fabricated high-precision polycrystalline diamond tools. *Int. J. Mach. Tools Manuf.*, 65, 99–106.

Zhao, W.S., Jia, B.X., Wang, Z.L., Hu, F.Q., 2006. Study on block electrode discharge grinding of micro rods. *Key Eng. Mater.* 304–305, 201–205.

7

Effect of Different Machining Parameters on the Quality of Machining

7.1 Introduction

Two critical elements of micro electro discharge machining (EDM) that influence most of the machining performance parameters are discharge energy and dielectric flushing. As the discharge energy increases, the heat transferred to the workpiece and tool increases. As a result, the crater volume increases, machining time reduces, material removal rate (MRR) increases, tool wear rate (TWR) increases, surface roughness increases, and debris concentration in the dielectric increases. If the dielectric flushing method and flow pressure are inadequate to remove the debris from the machining zone, the percentage of shorting and arcing increases, machining time increases, and the material deposition on the tool electrode increases. As a result, MRR reduces and surface quality diminishes. Figure 7.1 shows the important parameters associated with micro EDM and the performance parameters used to measure the quality of machining.

7.2 Micro EDM Process Parameters

Discharge energy: Discharge energy is the measure of heat transferred to the electrodes. The method to calculate the discharge energy may vary according to the pulse generator used for power supply. In the transistor circuit, the duration of the current flow is controlled by the gate switch. Average values of discharge current and voltage can be used to calculate the discharge energy if the pulse duration is known. The discharge energy per pulse in a transistor-type pulse generator is given by Eq. (7.1).

$$E = VIt_{on}$$

(7.1)

where E is the discharge energy per pulse, V is the discharge voltage, I is the discharge current, and t_{on} is the pulse-ON period.

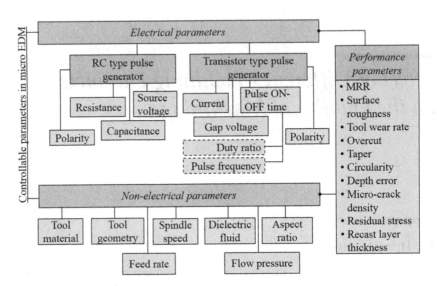

FIGURE 7.1
Machining parameters and performance parameters in micro EDM.

In a resistance-capacitance (RC) pulse generator, the capacitor is charged first (pulse-OFF time). Once the electrons overcome the potential barrier in the spark gap, the capacitor discharges the stored energy to the discharge gap. The discharge energy can be calculated from the discharge voltage (V) and capacitance (C) in the circuit, which is given by Eq. (7.2).

$$E = \frac{1}{2}CV^2$$

$$(7.2)$$

Discharge voltage: If V_g is the voltage across the spark gap and V is the discharge voltage in an RC circuit, the discharge happens in the gap when the magnitude of V_g reaches V. The voltage across the gap changes with respect to charging of capacitance. The discharge voltage in an RC circuit is given by Eq. (7.3). Here, R is the resistance and t_c is the cycle time (including charging and discharging of the capacitor).

$$V = V_0\left(1 - e^{\frac{-t_c}{RC}}\right)$$

$$(7.3)$$

Discharge current: Discharge current can be represented in terms of peak current and average current. Peak current is the maximum magnitude of current during machining. Average current is the average of discharge currents in a complete pulse cycle. Average current can be calculated by multiplying the peak current with duty factor (the ratio of pulse-ON time to total cycle time).

Pulse duration: It is the duration of the discharge current flow in the spark gap. When pulse duration increases, the plasma expands, and the heat is dissipated to the workpiece for a longer duration.

Pulse interval: This is the pulse-OFF time between two successive sparks, which is provided for deionization of the dielectric fluid and flushing of debris particles from the machining zone.

Polarity: In micro EDM, the workpiece is usually connected to the anode and the tool is connected to the cathode. The percentage of heat transferred to the workpiece depends on the polarity.

Tool rotation speed: It is the spindle speed during EDM drilling and milling process. Tool rotation speed has a direct influence on the efficiency of debris flushing as it creates more turbulence in the machining zone.

Tool electrode material: The electrical conductivity of the tool electrode will influence the breakdown voltage in micro EDM. The thermal conductivity of the electrode affects the heat transfer rate and material erosion.

Dielectric fluid: Deionized water and hydrocarbon oil-based dielectric fluids are commonly used in micro EDM. Recast layer formation, MRR, and TWR are affected by the type of dielectric fluid chosen for machining.

Tool feed: In micro EDM drilling, the speed of movement of the tool has to be controlled with respect to the MRR to avoid short-circuiting.

Tool geometry: The geometry of the tool will influence the dielectric flow in the machining zone. A notch or helical flute can be fabricated on the tool, which improves the debris removal efficiency.

Flushing pressure: Thin tool electrodes may deflect, vibrate, or bend if the flow pressure is high. If the flow pressure is low, debris removal becomes challenging as the working depth increases.

7.3 Performance Parameters

7.3.1 Material Removal Rate

It is defined as the rate of material erosion from the workpiece surface. MRR is calculated by measuring the weight of the workpiece before and after machining and dividing it with the machining time (g/min). It can also be quantified by calculating the ratio of machined volume and the time taken for material removal (mm^3/min). As the gap voltage (V), capacitance (C), and peak current (I) increase, the sparks carry more discharge energy. As the power and power density increase, more material from workpiece and tool is melted and flushed away, resulting in a higher MRR. MRR also depends on the type of pulse generator used in the system. As the transistor-type generator can produce high-frequency pulses, the MRR is observed to be higher (Jahan et al., 2009b).

According to Tiwary et al. (2014), as the peak current increases, the MRR shows an increasing trend. However, beyond a value of current, the MRR is

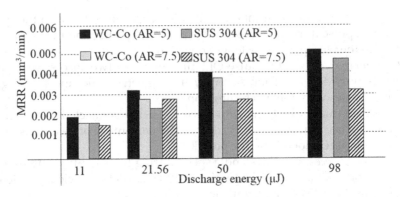

FIGURE 7.2
Effect of discharge energy on MRR (Jahan et al., 2010b).

observed to be decreasing. This may be caused by the increased tool wear and greater carbon deposition on the workpiece surface due to the high peak current. In the case of pulse duration, a similar trend is observed by Jahan et al. (2012). As the pulse-ON period increases, the MRR will increase initially. However, longer pulse durations will result in a higher amount of arcing due to increased debris concentration in the machining region. Pulse-OFF time has to be selected with care as it affects the material removal characteristics significantly. If the pulse interval is too long, then the MRR will steadily decrease. On the other hand, if the pulse interval is too short, the dielectric fluid in the machining zone will not get enough time to deionize, and the plasma column will be formed at the same location. According to Jahan et al. (2012), the highest MRR is recorded when the duty ratio was in the range of 0.4–0.5. Figure 7.2 shows the change in MRR with respect to discharge energy when different workpiece materials are machined with micro EDM (Jahan et al., 2010b).

MRR also depends on the properties of the workpiece. Increase in resistivity, density, and melting point of the workpiece will result in a low MRR (D'Urso et al., 2016). The work function of the electrode affects the ignition delay period and influence the total machining time. According to Jahan et al. (2009a), CuW tool gives more MRR compared to W electrodes in finishing micro EDM. This is attributed to the higher breakdown potential associated with W tool electrode due to its higher atomic mass and hardness (Jahan et al., 2009a).

Spindle speed during micro EDM has a significant effect on MRR when it goes from stationary condition to a moderate speed. However, a further increase in the spindle speed shows a negligible influence on MRR. As the rotational speed increases, the turbulence in the machining zone increases, which helps to carry the debris accumulations out of the spark gap. Researchers reported that the effect of turbulence disappears slowly when the rotational speed is further increased. Very high spindle speed is observed to be detrimental to machining in high-aspect-ratio drilling process as the chance of tool breakage increases (Jahan et al., 2012). In conventional milling process, the MRR increases with the tool feed. However, in micro EDM, if the tool feed does not match with the amount of volume removed per discharge, frequent short circuits can occur. As the tool pulls back during short-circuiting, the total machining time increases eventually. However, a very small tool feed will also result in higher machining time. The tool progression speed has to be selected carefully to realize stable machining (Jahan et al., 2012).

Selection of servo reference voltage and servo control strategy also affects the MRR significantly. Servo control is done to maintain the spark gap constant and reduce the percentage of harmful discharges in the spark gap. As the percentage of normal discharge increases with the help of an efficient servo control system, the machining time wasted in tool pullback can be avoided, and the MRR can be improved (Han et al., 2004). An improvement in workpiece erosion rate is achieved by employing planetary movement to the tool (Egashira et al., 2005). In vibration-assisted micro EDM, higher MRR is reported when the vibration frequency is increased. This is attributed to the increase in effective discharge ratio (ratio of normal discharges to total discharges) due to efficient debris removal (Tong et al., 2008).

Type of dielectric can also change the MRR characteristics. According to Kibria et al. (2010), during micro EDM of titanium alloys with kerosene, TiC compounds are formed on the workpiece surface due to decomposition of the hydrocarbon oil. The melting point of the carbide compound is higher than that of the titanium alloy, which reduces the material erosion rate. Addition of powder particles in dielectric fluid during micro EDM shows a positive effect in MRR (Jahan et al., 2010a; Kibria et al., 2010). The particles in the spark gap help to increase the spark gap and facilitate debris removal via better dielectric flushing. The increased spark gap also reduces the percentage of harmful discharges and the machining time will be shorter. However, the presence of metal additives causes a bridging effect in the speak gap and early discharge of the capacitor. The spark energy decreases due to this premature discharging (Kansal et al., 2007). Moreover, as the concentration of the powder in dielectric increases, a large part of the discharge energy will be carried away by additives. The energy loss results in low power density and plasma pressure, which may lead to low MRR. However, Kibria et al. (2010) reported a higher machining rate using B_4C-mixed kerosene if the pulse duration is extended.

7.3.2 Tool Wear

The material erosion from the tool during machining is called tool wear. Tool wear is commonly quantified by calculating the TWR and relative tool wear ratio. TWR is the eroded tool volume per unit time. Relative tool wear ratio stands for the ratio of rate of tool wear to the rate of workpiece erosion.

Similar to MRR, as the discharge energy increases, the tool wear becomes higher. The duration of discharge energy can be controlled by the pulse-ON time. According to Jahan et al. (2009a), at higher capacitance, electrode wear ratio decreases. However, this cannot be taken as a direct measure of the reduction of tool wear. At higher capacitance, the MRR takes the upper hand compared to TWR, and the effective wear ratio reduces. Tool wear ratio becomes higher at extreme peak current settings. As the current reaches a higher magnitude, debris concentration in the machining zone is increased. Due to this, the percentage of arc discharges is also increased. At high peak current settings, the thermal energy dissipated per unit area will be higher in the thin tool electrode, and the tool wear increases drastically (Tiwary et al., 2014). The effect of discharge energy on the relative electrode wear ratio (EWR) is shown in Figure 7.3.

As the thermal conductivity, melting point, and mass density of the tool material increase, the tool will experience less wear (D'Urso et al., 2016). According to Tsai and Masuzawa (2004), the corner wear becomes prominent in a tool electrode with less thermal conductivity. The lower thermal conductivity of W electrodes makes it more susceptible to tool edge rounding than AgW or CuW tools (Jahan et al., 2009a). Using different tool geometries for improving MRR and TWR is attempted by Pellicer et al. (2009). Even though fabricating notches on the tool surface for enhancing the dielectric flow helps to improve the MRR, the TWR also increases accordingly. The reduction of available material from the tool surface will increase the thermal energy per unit area in the tool electrode, and it erodes more quickly. Rotation of the tool decreases the relative tool wear ratio.

As the debris removal becomes more efficient with the introduction of tool rotation, the MRR will be higher than the TWR, which makes the relative wear ratio small (Jahan et al., 2012). Increase in tool feed increases the tool wear ratio as the arcing and shorting become high with high feed rate

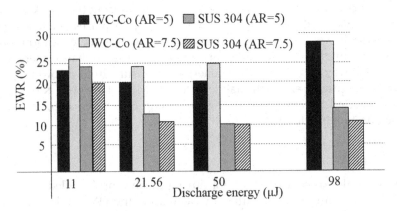

FIGURE 7.3
Effect of discharge energy on EWR during micro EDM (Jahan et al., 2010b).

(Jahan et al., 2012). The role of dielectric fluid in determining the TWR is discussed by Kibria et al. (2010). Micro EDM with kerosene as dielectric shows lesser tool wear than deionized water based machining. As the machining process progresses, the hydrocarbon oil decomposes and deposits a carbon layer on the tool surface. This carbon layer slows down the tool erosion process. According to Kibria et al. (2010), adding B_4C abrasive to the kerosene dielectric fluid helps to reduce the tool wear. The carbide particles also participate in the carbon deposition phenomena, which decelerate the rate of tool wear. As the aspect ratio increases, the MRR slows down and TWR increases, which results in a higher relative electrode wear ratio (Ali et al., 2009).

7.3.3 Overcut

Side erosion due to secondary discharges is the primary reason for overcut in micro EDM. Even though the debris particles are washed away from the primary machining zone, they can cause damage to the sidewalls as it floats in the fluid and causes side sparking. Increase in voltage and capacitance causes a high MRR, and the concentration of debris in the dielectric fluid increases. This may increase the frequency of secondary sparking and overcut. Moreover, at extreme voltage settings, the dielectric breakdown can happen even at wider spark gap (Jahan et al., 2009b). Additionally, a high concentration of debris particle in the dielectric fluid reduces the dielectric strength of the fluid, which may result in arcing and higher overcut. Overcut is observed to be less while using deionized water as a dielectric at low peak current settings (Kibria et al., 2010). Nonetheless, at a high peak current setting, kerosene behaves better in terms of overcut compared to deionized water. This may be due to the release of a large amount of oxygen from the decomposition of deionized water at the high current setting. The oxygen bubbles may lead to secondary discharges and hence causes a higher overcut. Adding B_4C particles to the dielectric increases overcut. Kibria et al. (2010) attributed this phenomenon to the removal of the recast layer in the sidewalls in the presence of abrasive additives in the fluid. The bridging phenomena between the powder additives and sidewalls will also cause an increased overcut (Kansal et al., 2007). Figure 7.4 shows the effect of discharge energy on overcut for different workpiece materials and aspect ratios.

According to D'Urso et al. (2012), an increase in tool rotation speed results in high overcut. This may be due to the enhanced removal of molten material from the side surface because of higher turbulence in the fluid. Overcut will also depend on the workpiece material (Jahan et al., 2010b). Hole drilled in stainless steel (SS 304) material shows higher overcut values when compared to WC-Co material at high discharge energy settings as well as for high-aspect-ratio holes. As the thermal expansion coefficient of SUS 304 is three times higher than WC-Co, the increase in overcut can be due to thermal expansion of the material during deep-hole drilling with high discharge energy. When the aspect ratio increases, debris removal will become challenging, and

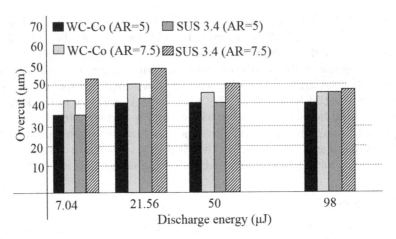

FIGURE 7.4
Effect of discharge energy on overcut (Jahan et al., 2010b).

secondary discharges can cause expansion of the hole (Jahan et al., 2010b). The overcut values dropped as the orbital radius in planetary EDM increased. This may be due to better control of the side surface machining as the cylindrical part of the tool electrode participates more in the drilling process (Bamberg and Heamawatanachai, 2008). As the debris removal is improved by giving a tool or workpiece vibration, the overcut values drop with an increase in vibrational frequency. However, at very high vibration frequency, the tool side surface may touch the hole sidewalls frequently, which leads to more harmful discharges and high overcut (Jahan et al., 2012).

7.3.4 Taperness

Corner wear of the tool electrode leads to taper in the microholes. As the voltage and capacitance increase, the increase in tool wear will be reflected as taperness in the microfeatures. According to Kumar et al. (2018), the trend concerning peak current changes with the tool electrode composition. For the Cu electrode, the taper rate is decreased when the peak current is in the range of 2–4 A. However, a reverse trend is observed when the peak current increased to 4–7 A. Interestingly for brass, taper decreases when the peak current increased from 3 to 8 A. The difference in thermal conductivity is suggested as the reason behind the variation in taperness. Increase in secondary discharges at the sidewalls with an increase in peak current causes hole expansion, which is suggested as the reason behind the reduction of taper by Pradhan et al. (2009). Figure 7.5 shows the change in taper angle with respect to discharge energy during micro EDM of various workpiece materials. Taperness seems to be higher in WC-Co compared to SUS 304 (Jahan et al., 2010b).

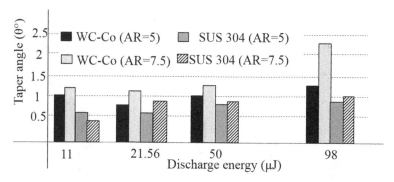

FIGURE 7.5
Effect of discharge energy on taper angle (Jahan et al., 2010b).

As pulse duration increases, the taper increases. However, at high current settings, increase in pulse-on time results in the formation of a carbonaceous layer on the tool electrode, and the taper reduces due to reduced tool wear. As the stability of machining increases with an increase in pulse interval, the taper is decreased at the high pulse-OFF time (Kumar et al., 2018). The average taper angle is observed to be smaller with the RC-type pulse generator than transistor type (Jahan et al., 2009b). Addition of powder particles to the dielectric also reduces the taper (Kibria et al., 2010). The taper increases with depth due to debris accumulation in high-aspect-ratio holes (Ali et al., 2009). According to Feng et al. (2018), very high tool rotation speed results in straight wall microfeatures.

7.3.5 Circularity and Depth Error

Lateral tool vibrations can be the main reason for roundness error in micro EDM drilling (Kadirvel et al., 2013). As the aspect ratio increases, the thin tool may be subjected to large lateral vibrations, which leads to an increase in roundness error (Ali et al., 2009). According to Jahan et al. (2009b), employing RC-type circuits in micro EDM drilling resulted in holes with less circularity errors. The depth error can be considered as a function of tool wear and MRR in micro EDM. The effect of parameters on the depth error can be assessed by analyzing the tool wear and workpiece erosion trends. Figure 7.6 shows the trend of roundness error with respect to discharge energy.

7.3.6 Short Circuit

With low discharge energy, the material removal will be slow, and the feeding of the tool forward will result in frequent short circuits. As the voltage and capacitance increases, the removed crater depth will be high, and the spark gap increases. This will reduce the chance of short-circuiting in micro EDM. However, as the MRR increases, the machining zone will be filled

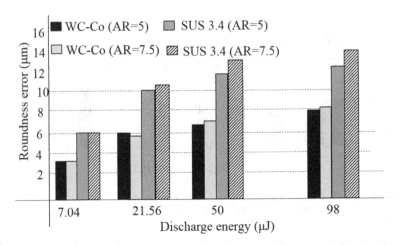

FIGURE 7.6
Effect of discharge energy in circularity error (Jahan et al., 2010b).

with more and more debris particles. The higher concentration of debris in the spark gap will increase the number of short circuits. If the feed rate does not match with the MRR, then the chance of frequent shorting increases considerably (Jahan et al., 2010a). The servo reference voltage also plays a vital role in controlling shorting during micro EDM. Efficient pulse discriminators and optimum servo reference voltage can reduce the chance of contact between the tool electrode and workpiece during machining (Yang et al., 2011). As the aspect ratio increases, the debris entrapment becomes significant leading to frequent shorting (Ali et al., 2009). Tool rotation creates a turbulence in the machining region, removes debris, and encourages stable discharges. Reducing the pulse interval time will result in massive arcing and short-circuiting due to a reduction in time for deionizing the dielectric fluid (Jahan et al., 2012). An increase in spark gap is observed with the addition of powder particles to the dielectric fluid (Klocke et al., 2004). The increase in vibrational frequency in micro EDM results in less shorting and stable machining (Endo et al., 2008). However, extreme vibrations to the tool will cause more shorting as the tool may touch the sidewalls during machining. Figure 7.7 shows the change in the percentage of short-circuit pulses with respect to a change in discharge energy.

7.3.7 Surface Roughness

As the discharge energy goes up, the machined crater becomes deeper. The valley-to-peak distance in the machined surface increases, which leads to high surface roughness. Moreover, in the RC pulse generator, the nonuniformity of crater dimensions due to nonuniform pulse energy reduces the quality of the machined surface. According to Jahan et al. (2009b), surface

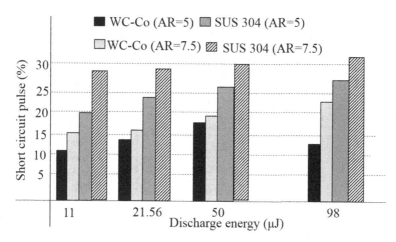

FIGURE 7.7
The effect of discharge energy on percentage of short-circuit pulses (Jahan et al., 2010b).

machined with RC pulsed micro EDM shows less surface roughness but never gets a glossy surface. The uneven crater distribution in the surface is known to be the reason for this. According to them, the glossiness increases with uniform overlapping craters. The uniformity in crater distribution can be influenced by the thermal and electrical conductivity of the electrode material too. Higher electrical conductivity offers uniform spark distribution over the workpiece surface (Jahan et al., 2009a). As the pulse duration increases, the plasma channel expands largely, which reduces the power density on the workpiece surface. This will result in wide and shallow craters (Mujumdar et al., 2015). Increase in pulse frequency results in more overlapping craters and a reduction in surface roughness (Jahan et al., 2009a). It is obvious that the surface roughness increases with an increase in voltage and capacitance. However, machining at very low voltage produces a rough surface. This is attributed to poor machining (arcing and short-circuiting) at a lower voltage range (Jahan et al., 2009a). Recast layer and microcracks on the workpiece surface also increase the average roughness value (Ekmekci et al., 2009). Adding orbital movement to the tool reduces the surface roughness compared to the normal micro EDM process (Bamberg and Heamawatanachai, 2008).

7.3.8 Recast Layer Thickness

When the discharge current is increased, the recast layer thickness becomes uneven. The high pulse current results in a thick molten pool with small crater radius. Due to high-pressure plasma, the molten layer flows to the rim of the crater and solidifies to form burr-like recast layer on the surface (Ekmekci et al., 2009). This spilling of molten layer to the surface increases the deviation in recast layer thickness over the surface. However, when the pulse duration is long, more shallow craters with a thin molten layer and

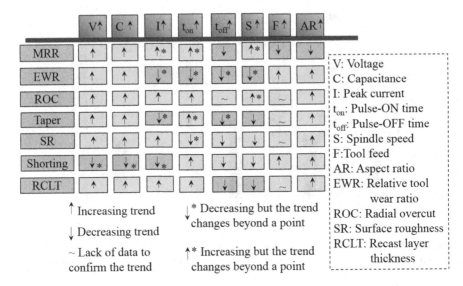

FIGURE 7.8
Machining parameters in micro EDM and their influence on performance parameters.

large crater radius are formed. The increased interaction time of the molten layer with the hydrocarbon dielectric fluid will result in carbon contamination in the recast layer (Ekmekci et al., 2009). Overlapping recast layers will result in microcrack formation. Maximum crack density is observed at maximum pulse-ON time settings. As the pulse-ON time increases, the carbon deposition in the recast layer increases due to prolonged interaction between the hydrocarbon-based dielectric fluid and molten pool (Ekmekci et al., 2009). The tensile stress in the surface layer and subsequent layers will result in crack formation. Figure 7.8 summarizes the effect of each machining parameter on the performance parameter during micro EDM.

7.4 Conclusion

Process parameters are categorized into electrical parameters (discharge energy, current, capacitance, voltage, pulse duration, pulse interval, polarity, etc.) and machining parameters (tool rotation, flushing pressure, feed rate, tool–workpiece material combination, etc.). Magnitude and combination of these parameters significantly affect the quality of the machined surface. Aggressive setting of these parameters results in low processing time but high dimensional error and vice versa. So, optimum selection and combination are required to balance the MRR and high dimensional accuracy.

References

Ali, M.Y., Hamad, M.H., Karim, A.I., 2009. Form characterization of microhole produced by microelectrical discharge drilling. *Mater. Manuf. Process.* 24, 683–687.

Bamberg, E., Heamawatanachai, S., 2008. Orbital electrode actuation to improve the efficiency of drilling micro-holes by micro-EDM. *J. Mater. Process. Technol.* 209, 1826–1834.

D'Urso, G., Longo, M., Maccarini, G., Ravasio, C., 2012. Electrical discharge machining of micro holes on titanium sheets. *Proc. ASME Des. Eng. Tech. Conf.* 7, 417–424.

D'Urso, G., Maccarini, G., Ravasio, C., 2016. Influence of electrode material in micro-EDM drilling of stainless steel and tungsten carbide. *Int. J. Adv. Manuf. Technol.* 85, 2013–2025.

Egashira, K., Taniguchi, T., Hanajima, S., 2005. Planetary EDM of micro holes. *Ratio* 11, 15–18.

Ekmekci, B., Sayar, A., Öpöz, T.T., Erden, A., 2009. Geometry and surface damage in micro electrical discharge machining of micro-holes. *J. Micromech. Microeng.* 19, 105030.

Endo, T., Tsujimoto, T., Mitsui, K., 2008. Study of vibration-assisted micro-EDM-the effect of vibration on machining time and stability of discharge. *Precis. Eng.* 32, 269–277.

Feng, G., Yang, X., Chi, G., 2018. Experimental and simulation study on micro hole machining in EDM with high-speed tool electrode rotation. *Int. J. Adv. Manuf. Technol.* 101(1–4), 367–375.

Han, F., Wachi, S., Kunieda, M., 2004. Improvement of machining characteristics of micro-EDM using transistor type isopulse generator and servo feed control. *Precis. Eng.* 28, 378–385.

Jahan, M.P., Rahman, M., Wong, Y.S., 2010a. Modelling and experimental investigation on the effect of nanopowder-mixed dielectric in micro-electrodischarge machining of tungsten carbide. *Proc. Inst. Mech. Eng. Part B J. Eng. Manuf.* 224, 1725–1739.

Jahan, M.P., San Wong, Y., Rahman, M., 2010b. A comparative experimental investigation of deep-hole micro-EDM drilling capability for cemented carbide (WC-Co) against austenitic stainless steel (SUS 304). *Int. J. Adv. Manuf. Technol.* 46, 1145–1160.

Jahan, M.P., Wong, Y.S., Rahman, M., 2009a. A study on the fine-finish die-sinking micro-EDM of tungsten carbide using different electrode materials. *J. Mater. Process. Technol.* 209, 3956–3967.

Jahan, M.P., Wong, Y.S., Rahman, M., 2009b. A study on the quality micro-hole machining of tungsten carbide by micro-EDM process using transistor and RC-type pulse generator. *J. Mater. Process. Technol.* 209, 1706–1716. doi:10.1016/j.jmatprotec.2008.04.029.

Jahan, M.P., Wong, Y.S., Rahman, M., 2012. Experimental investigations into the influence of major operating parameters during micro-electro discharge drilling of cemented carbide. *Mach. Sci. Technol.* 16, 131–156.

Kadirvel, A., Hariharan, P., Gowri, S., 2013. Experimental investigation on the electrode specific performance in micro-EDM of die-steel. *Mater. Manuf. Process.* 28, 390–396.

Kansal, H.K., Singh, S., Kumar, P., 2007. Technology and research developments in powder mixed electric discharge machining (PMEDM). *J. Mater. Process. Technol.* 184, 32–41.

Kibria, G., Sarkar, B.R., Pradhan, B.B., Bhattacharyya, B., 2010. Comparative study of different dielectrics for micro-EDM performance during microhole machining of Ti-6Al-4V alloy. *Int. J. Adv. Manuf. Technol.* 48, 557–570.

Klocke, F., Lung, D., Antonoglou, G., Thomaidis, D., 2004. The effects of powder suspended dielectrics on the thermal influenced zone by electrodischarge machining with small discharge energies. *J. Mater. Process. Technol.* 149, 191–197.

Kumar, K., Kumar Rawal, S., Singh, V.P., Bala, A., 2018. Experimental study on diametric expansion and taper rate in EDM drilling for high aspect ratio micro holes in high strength materials. *Mater. Today Proc.* 5, 7363–7372.

Mujumdar, S.S., Curreli, D., Kapoor, S.G., Ruzic, D., 2015. Model-based prediction of plasma resistance, and discharge voltage and current waveforms in microelectrodischarge machining. *J. Micro Nano Manuf.* 4, 011003.

Pellicer, N., Ciurana, J., Ozel, T., 2009. Influence of process parameters and electrode geometry on feature micro-accuracy in electro discharge machining of tool steel. *Mater. Manuf. Process.* 24, 1282–1289.

Pradhan, B.B., Masanta, M., Sarkar, B.R., Bhattacharyya, B., 2009. Investigation of electro-discharge micro-machining of titanium super alloy. *Int. J. Adv. Manuf. Technol.* 41, 1094–1106.

Tiwary, A.P., Pradhan, B.B., Bhattacharyya, B., 2014. Study on the influence of micro-EDM process parameters during machining of Ti–6Al–4V superalloy. *Int. J. Adv. Manuf. Technol.* 76, 151–160.

Tong, H., Li, Y., Wang, Y., 2008. Experimental research on vibration assisted EDM of micro-structures with non-circular cross-section. *J. Mater. Process. Technol.* 208, 289–298.

Tsai, Y.-Y., Masuzawa, T., 2004. An index to evaluate the wear resistance of the electrode in micro-EDM. *J. Mater. Process. Technol.* 149, 304–309.

Yang, G.Z., Liu, F., Lin, H.B., 2011. Research on an embedded servo control system of micro-EDM. *Appl. Mech. Mater.* 120, 573–577.

8

Different Machining Inaccuracies and Their Causes

8.1 Introduction

Micro electro discharge machining (EDM) is recognized as a prominent machining technique because of some peculiar machining characteristics including force-free machining, the capability of 3D machining, the capability of fabricating thin microfeatures, etc. However, surface quality and dimensional accuracy are often affected by tool wear, secondary discharges, inefficient dielectric flushing, etc. These irregularities can be broadly classified into two sections: geometrical errors and surface irregularities. Geometrical errors mainly occur due to tool electrode wear and side erosion due to floating debris particles. Causes for surface irregularities include temperature gradient in the workpiece, resolidification of molten material, burning of dielectric fluid, and debris accumulation on the tool surface. General classification of the errors associated with micro EDM is shown in Figure 8.1.

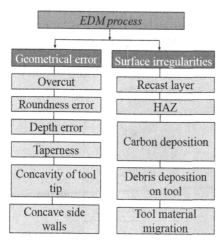

FIGURE 8.1
Different geometrical errors and surface irregularities in micro EDM.

8.2 Geometrical Errors

The errors related to micro EDM responsible for dimensional changes and shape inaccuracy are brought under this classification.

8.2.1 Overcut

Overcut is defined as the difference in the intended diameter of the hole and the actual diameter after machining. Discharges from the tool side surface remove extra material from the sidewalls of the holes, which results in expansion of the holes. Machining parameters such as discharge energy and pulse duration influence the amount of expansion (Kadirvel et al., 2013). Overcut is measured as the difference between the tool diameter (D_e) and the hole diameter (D_h), as shown in Figure 8.2. The progression of overcut with respect to increasing aspect ratio is shown in Figure 8.3.

8.2.2 Taperness

Side wear in tool electrode, as well as the secondary discharges away from the machining zone, makes the hole or channel cross section tapered. The floating debris particles cause secondary discharges at different locations away from the primary machining zone. This makes the cross section of holes nonuniform. In micro EDM drilling, the taper is measured as a function of the difference between the entrance and exit diameters and the hole depth. Figure 8.4 shows the change in hole diameter at the entrance and exit sections of a microhole drilled with EDM (Feng et al., 2019). The taperness may also depend on the machining parameters and type of pulse generator used for machining. According to Jahan et al. (2009), resistor-capacitor (RC)-type pulse generator produces holes with less taper. Taper angle (α) in micro EDM drilling can be calculated using Eq. (8.1). Figure 8.5a shows a schematic

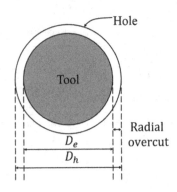

FIGURE 8.2
Schematic representation of radial overcut in micro EDM.

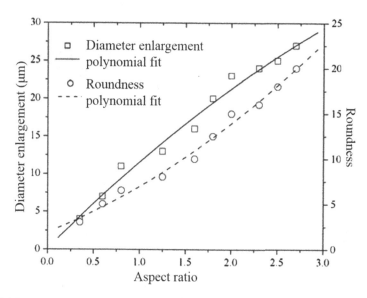

FIGURE 8.3
Relationship between roundness error and overcut with the aspect ratio during micro EDM drilling (Ali et al., 2009).

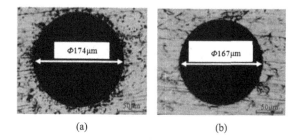

(a)   (b)

FIGURE 8.4
Diameter of the microhole drilled with EDM at (a) entrance section (174 µm) and (b) exit section (167 µm) (Feng et al., 2019).

representation of parameters used in taper angle calculation, where $D_{h_{ent}}$ is the entrance diameter of the hole, $D_{h_{ext}}$ is the exit diameter of the hole, and Z is the hole depth. Figure 8.5b shows the taper angle in a blind hole drilled using micro EDM drilling.

$$\alpha = \tan^{-1}\left(\frac{D_{h_{ent}} - D_{h_{ext}}}{Z}\right) \tag{8.1}$$

The effect of aspect ratio on the taper angle of the hole is studied by Ali et al. (2009). The experimental results are curve fitted to formulate an empirical relationship between taper angle and aspect ratio. The taper angle increases

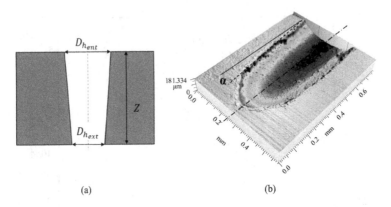

(a)   (b)

FIGURE 8.5
Taper in microhole drilled with EDM: (a) schematic diagram of microhole geometry and (b) 3D profile of a blind microhole showing the taper angle.

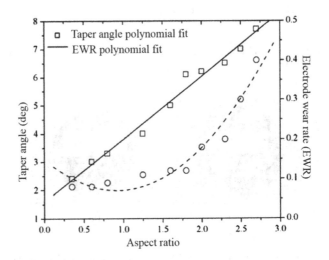

FIGURE 8.6
Effect of aspect ratio on taper angle during micro EDM drilling (Ali et al., 2009).

with hole depth because of increased tool wear and increase in the percentage of secondary discharges (Figure 8.6).

8.2.3 Roundness Error

During micro EDM drilling, the vibration of the tool electrode results in circularity errors (Figure 8.7). The amount of roundness error is influenced by the rotational speed of the tool and the type of pulse generator used. RC-type pulse generator has the advantage of producing holes with less roundness error (Jahan et al., 2009). The roundness error can be quantified

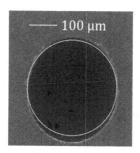

FIGURE 8.7
Change in the roundness of a microhole machined with EDM (Ali et al., 2009).

by various parameters including least square circle, minimum zone circle, maximum inscribed circle, and minimum circumscribed circle (Kadirvel et al., 2013). According to Kadirvel et al. (2013), the roundness error also depends on the type of electrode used for machining. Machining at lower voltage range using Cu and CuW electrodes shows an increase in roundness error. W electrodes exhibit comparatively smaller circularity error at low voltage. Elevated roundness error at lower voltage range is attributed to poor machining. At higher voltage range, roundness error increases due to material erosion due to arcing (Kadirvel et al., 2013). The relationship of roundness error with the aspect ratio is studied by Ali et al. (2009) as shown in Figure 8.3. The error is found to increase with hole depth. To drill deeper holes, a long tool has to be used. Employing long microrods leads to extensive vibration during drilling and results in nonsymmetrical cross section of the machined hole. Fluctuations in the power supply also cause detrimental effects on the structural uniformity of the hole (Ali et al., 2009).

8.2.4 Bottom Rounding

Corner wear of the tool electrode will also result in edge rounding of the holes and channels as shown in Figure 8.8a. As the corner wear ratio increases, the rounding error will also increase as shown in Figure 8.8b, where the radius of curvature R2 is greater than R1.

8.2.5 Depth Error

Due to tool wear, the depth of microchannel during micro EDM milling, the height of electrode array during reverse EDM, and the depth of hole during micro EDM drilling are reduced from the expected value (Guo et al., 2006; Yeo et al., 2007; Mastud et al., 2012). Depth errors can be tackled by implementing appropriate tool compensation techniques (Dimov et al., 2003; Pham et al., 2007). Figure 8.9 shows the change in hole depth during the drilling of a series of holes without employing any tool compensation methods (Malayath et al., 2019).

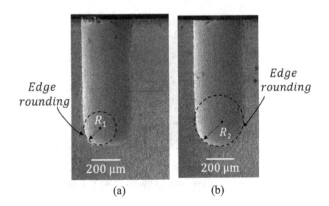

FIGURE 8.8
(a) and (b) Increase in edge rounding with respect to increase in the corner wear of the tool.

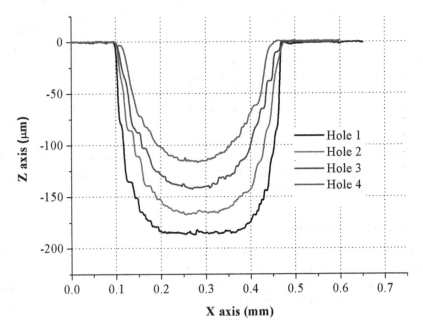

FIGURE 8.9
Change in hole depth during drilling of a series of holes (Malayath et al., 2019).

8.2.6 Width Error

Corner wear during micro EDM milling will result in non-uniform width as the machining progresses as shown in Figure 8.10 (Hung et al., 2011). To keep the channel width constant, techniques like layer-by-layer machining or periodical flattening of the tool bottom surface have to be employed (Karthikeyan et al., 2011).

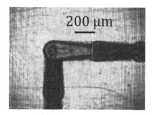

FIGURE 8.10
Change in channel width during micro EDM milling (Hung et al., 2011).

8.2.7 The Concavity of Tool Tip and Debris Accumulation at the Hole Bottom

During machining with low voltage and short durations in micro EDM drilling, the tool electrode tip is often transformed into a concave shape. During blind hole machining, the molten material ejected from the crater rotates with the vortex below the tool electrode due to tool rotation. The debris particles are cooled down and accumulated at the tool bottom surface when the dielectric fluid fails to wash away the particles. The vortex motion causes a heap of debris particles at the hole bottom (Ekmekci and Sayar, 2013). Discharges between the accumulated debris and the tool electrode eventually remove material from the tool electrode (Figure 8.11a). Finally, a concave tool tip and protrusion at the hole bottom are produced (Figure 8.11b). When the tool is not rotated, the protrusion at the hole bottom has a nonuniform shape (Ekmekci and Sayar, 2013).

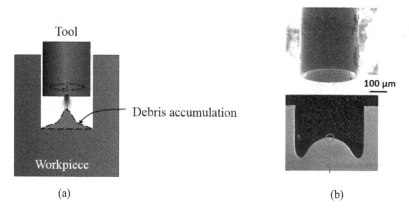

FIGURE 8.11
(a) Debris accumulation at the hole bottom and formation of discharges and (b) microelectrodes and microholes drilled with low-voltage short pulses (Ekmekci and Sayar, 2013).

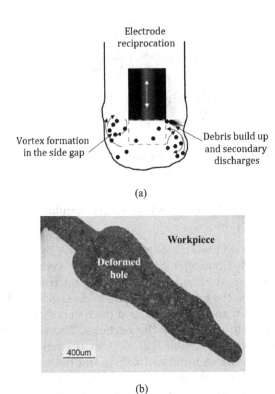

(a)

(b)

FIGURE 8.12
(a) Vortex formation in the side gap during die-sinking micro EDM and redistribution of debris and (b) deformed hole with concave walls (Murray et al., 2012).

8.2.8 Concave Walls

During die-sinking EDM, the reciprocating electrode makes a vortex in the side gap, and the debris particles are redistributed (Murray et al., 2012). Secondary discharges from the debris flow create concave walls during machining as explained in Figure 8.12a. Figure 8.12b shows a cross section of a microhole with concave walls. These deformations severely affect the shape accuracy of the machined microfeature.

8.3 Surface Irregularities

Irregularities on the surface machined with micro EDM is attributed to the heat flow from the discharges, formation of recast layer, and contamination of the recast layer. Thermally influenced layers affect the metallurgical

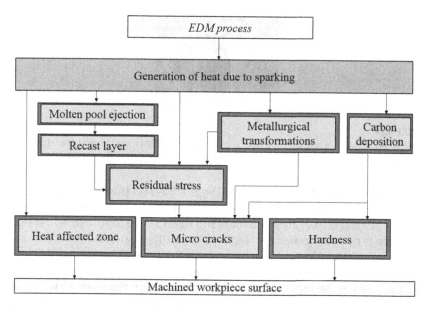

FIGURE 8.13
Effect of heat on the quality of machined surface in micro EDM.

and mechanical properties of the workpiece material. Formation of surface irregularities due to temperature variations during micro EDM is summarized in Figure 8.13.

Thermally influenced sublayers in the surface machined by micro EDM can be classified according to the hardness and microstructure of the material in a specific layer (Kadirvel et al., 2013). The thickness and material structure of these layers depend on the material composition and thermal conductivity of the workpiece. The first layer on the top, recognized by the loosely bounded material that can be washed away with pressurized dielectric flushing, is known as the spattered debris layer. Beneath which a layer of deposited material formed due to resolidification of the molten material is called the recast layer or white layer. Finally, the layer in which the properties are influenced by the heat conducted from the machining zone is called the heat-affected zone (HAZ).

8.3.1 Recast Layer

A part of the molten material thrown out from the discharge crater will stick to the crater rim and cool down rapidly during the pulse-off period. This layer is recognized by elevated hardness, brittleness, and porosity. As the machining progresses, the recast layer thickness increases due to the addition of more molten material on the existing layer. This overlapping of recast layers will result in the formation of residual stress and cracking on the layer.

FIGURE 8.14
Burr-like recast layer formed on the rim of the microhole fabricated with a transistor pulse generator (Jahan et al., 2009).

Formation of a burr-like recast layer (Figure 8.14) on the entrance section of the microhole is visible in EDM drilling (Ekmekci et al., 2009). According to Jahan et al. (2009), burr formation during EDM is predominant while using transistor-type pulse generator. Tan and Yeo (2011) quantified the recast layer thickness by estimating average recast layer thickness. The average thickness is calculated by dividing the machined surface into vertical elemental strips. The total area of the recast layer is then divided with the element width and number of sections. The effect of tool electrode rotation and the presence of powder particles in the dielectric fluid on the recast layer thickness are analyzed by Tan and Yeo (2011). Electrode rotation reduces the resolidification of molten material by enhancing the dielectric flow. However, the presence of powder particles causes a reduction in plasma pressure, which results in ineffective molten pool flushing and formation of overlapping recast layers.

8.3.2 Residual Stress

During solidification of the molten material, the shrinkage is restrained by the cold layer beneath it. As a result, tensile stresses are generated in the new layer (Bleys et al., 2006). The formation of tensile stresses in the resolidified layer weakens the effective strength of microfeatures. The residual stress generated during the fabrication of microelectrodes using wire EDG causes side bending to the rods. The amount of bending depends on the mechanical properties of the electrode material (Kawakami and Kunieda, 2005).

8.3.3 Microcracks

The formation of microcracks is attributed to the rapid cooling of the molten material. In steel, high quenching rates can cause the formation of an amorphous structure in the recast layer with a martensitic structure. Interestingly, studies conducted by Bleys et al. (2006) revealed that the martensitic structure is only visible in a small number of distributed spots along the recast layer. Figure 8.15 shows the change in workpiece surface characteristics after machining. Microcracks are visible all over the surface. Additionally, micropores are also found on the workpiece surface. Chen et al. (2014) showed a method to reduce microcracks by adding Ti powder in the dielectric fluid.

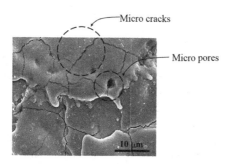

FIGURE 8.15
Workpiece surface after micro EDM (Chen et al., 2014).

8.3.4 Heat-Affected Zone

The area affected by the heat generated from discharges is called the "heat-affected zone". The thermal influence of the discharges changes the mechanical properties of the layer. As the discharge energy is small in micro EDM, distinguishing the HAZ will not be easy for all materials. However, HAZ in steel is often recognized by the presence of a hardened (martensitic) and annealed layer (Bleys et al., 2006). The HAZ area depends on the discharge intensity (Ekmekci et al., 2009) and the thermal diffusivity of the material. HAZ is deeper when machining with Cu, Cu-W than W and Ag electrodes. According to Kadirvel et al. (2013), machining with tungsten electrodes provides very shallow HAZ and negligibly small recast layer. Thao and Joshi (2008) studied the micro EDM drilling process to analyze the geometrical imperfections during machining and reported the presence of HAZ with a thickness in the range of 30–100 μm. Machining carbon fiber reinforced plastic with micro EDM shows different HAZ characteristics. The low thermal conductivity of the epoxy resin resulted in early melting of the material and flushed away easily if a high discharge current is used. This results in delamination, and hence, the surface quality will be severely compromised (Teicher et al., 2013). Pulse duration also plays an important role in determining the thickness of thermally influenced layers as more heat sinks into the workpiece surface as the discharge time increases.

8.3.5 Tool Electrode Migration

Material transfer from the tool to the workpiece is observed during micro EDM, which affects the composition of the workpiece material. The migration of tool material to the workpiece surface also affects the mechanical properties of the surface and subsurface layers (Jahan et al., 2010). According to Bleys et al. (2006), the thickness of Cu deposition found on the surface of steel workpiece during milling EDM was higher than the depositions during die sinking. Most of the contamination due to tool electrode material is found on the recast layer.

8.3.6 Carbon Deposition

Addition to the formation of thermally influenced layers, dispersive X-ray spectrometry analysis revealed the presence of carbon depositions on the workpiece surface (Ekmekci et al., 2009). Pyrolysis of hydrocarbon-based dielectric during EDM is responsible for carbon deposition on the workpiece surface. Most of the carbon deposition is located in the recast layer, which increases the hardness of the layer in certain materials like steel. In steel, deposition of carbon causes formation of iron carbide compounds (e.g., Fe3C), which increases the hardness of the recast layer (Bleys et al., 2006).

8.3.7 Depositions on the Tool

The tool is often contaminated by workpiece material depositions. According to Murray et al. (2012), workpiece-rich layer is present on the tool electrode during micro EDM for all range of machining parameters. Interestingly, the workpiece deposition is found at different locations away from the primary machining zone. The remelting of debris particles due to secondary discharges and reattachment of the molten material on the tool surface are considered to be the causes of workpiece deposition on the tool surface.

8.3.8 Electrolytic Corrosion

When deionized water is chosen as a dielectric fluid, the machined surface quality is adversely affected by electrolytic corrosion. For example, during micro EDM of WC-Co alloy, cobalt is electrolytically dissolved in the dielectric fluid due to electrolyzation of cobalt in the positively charged workpiece as shown in Figure 8.16. The electrolytic reactions deteriorate the quality of the machined surface as shown in Figure 8.17.

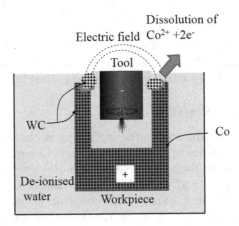

FIGURE 8.16
Electrolytic corrosion during micro EDM using deionized water as the dielectric fluid.

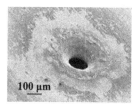

FIGURE 8.17
Effect of electrolytic corrosion on the surface quality of the microhole during machining, with deionized water as dielectric fluid (Chung et al., 2011).

8.4 Conclusion

Due to tool wear, debris accumulation, and temperature variations, various geometrical errors and surface irregularities are formed on the microfeature. The main reason behind the dimensional inaccuracy is tool wear (linear wear and corner wear). Debris accumulation will also result in shape distortions such as concave sidewalls and concave tool bottom. Some of the molten material thrown out from the crater resolidifies on the workpiece surface (recast layer). Overlapping of recast layers and difference in their solidification rates will result in the formation of tensile stress in the material. The tensile stress formation leads to the formation of microcracks and micropores. The recast layer is also contaminated by tool material migration and carbon deposition. When deionized water is used for machining, the surface quality is affected by electrolytic corrosion.

References

Ali, M.Y., Hamad, M.H., Karim, A.I., 2009. Form characterization of microhole produced by microelectrical discharge drilling. *Mater. Manuf. Process.* 24, 683–687.

Bleys, P., Kruth, J., Lauwers, B., Schacht, B., Balasubramanian, V., Froyen, L., Van Humbeeck, J., 2006. Surface and sub-surface quality of steel after EDM. *Adv. Eng. Mater.* 8, 15–25.

Chen, S.L., Lin, M.H., Huang, G.X., Wang, C.C., 2014. Research of the recast layer on implant surface modified by micro-current electrical discharge machining using deionized water mixed with titanium powder as dielectric solvent. *Appl. Surf. Sci.* 311, 47–53.

Chung, D.K., Shin, H.S., Park, M.S., Chu, C.N., 2011. Machining characteristics of micro EDM in water using high frequency bipolar pulse. *Int. J. Precis. Eng. Manuf.* 12, 195–201.

Dimov, S., Pham, D.T., Ivanov, A., Popov, K., 2003. Tool-path generation system for micro-electro discharge machining milling. *Proc. Inst. Mech. Eng. Part B J. Eng. Manuf.* 217, 1633–1637.

Ekmekci, B., Sayar, A., 2013. Debris and consequences in micro electric discharge machining of micro-holes. *Int. J. Mach. Tools Manuf.* 65, 58–67. doi: 10.1016/j.ijmachtools.2012.10.003.

Ekmekci, B., Sayar, A., Öpöz, T.T., Erden, A., 2009. Geometry and surface damage in micro electrical discharge machining of micro-holes. *J. Micromech. Microeng.* 19, 105030.

Feng, G., Yang, X., Chi, G., 2019. Experimental and simulation study on micro hole machining in EDM with high-speed tool electrode rotation. *The Int. J. of Adv. Manuf. Tech.* 101(1–4), 367–375.

Guo, R., Zhao, W., Li, G., Li, Z., Zhang, Y., 2006. A machine vision system for micro-EDM based on linux. *Third International Symposium on Precision Mechanical Measurements. International Society for Optics and Photonics*, Urumqi, China, 62803K.

Hung, J.-C., Yang, T.-C., Li, K., 2011. Studies on the fabrication of metallic bipolar plates—using micro electrical discharge machining milling. *J. Power Sources* 196, 2070–2074.

Jahan, M.P., Rahman, M., Wong, Y.S., 2010. Migration of materials during finishing micro-EDM of tungsten carbide. *Key Eng. Mater.* 443, 681–686.

Jahan, M.P., Wong, Y.S., Rahman, M., 2009. A study on the quality micro-hole machining of tungsten carbide by micro-EDM process using transistor and RC-type pulse generator. *J. Mater. Process. Technol.* 209, 1706–1716.

Kadirvel, A., Hariharan, P., Gowri, S., 2013. Experimental investigation on the electrode specific performance in micro-EDM of die-steel. *Mater. Manuf. Process.* 28, 390–396.

Karthikeyan, G., Sambhav, K., Ramkumar, J., Dhamodaran, S., 2011. Simulation and experimental realization of μ-channels using a μeD-milling process. *Proc. Inst. Mech. Eng. Part B J. Eng. Manuf.* 225, 2206–2219.

Kawakami, T., Kunieda, M., 2005. Study on factors determining limits of minimum machinable size in micro EDM. *CIRP Ann.* 54, 167–170.

Malayath, G., Katta, S., Sidpara, A.M., Deb, S., 2019. Length-wise tool wear compensation for micro electric discharge drilling of blind holes. *Meas. J. Int. Meas. Confed.* 134, 1–946.

Mastud, S., Singh, R.K., Joshi, S.S., 2012. Analysis of fabrication of arrayed micro-rods on tungsten carbide using reverse micro-EDM. *Int. J. Manuf. Technol. Manag.* 26, 176–195.

Murray, J., Zdebski, D., Clare, A.T., 2012. Workpiece debris deposition on tool electrodes and secondary discharge phenomena in micro-EDM. *J. Mater. Process. Technol.* 212, 1537–1547.

Pham, D.T., Ivanov, A., Bigot, S., Popov, K., Dimov, S., 2007. An investigation of tube and rod electrode wear in micro EDM drilling. *Int. J. Adv. Manuf. Technol.* 33, 103–109.

Tan, P.C., Yeo, S.H., 2011. Investigation of recast layers generated by a powder-mixed dielectric micro electrical discharge machining processg. *Proc. Inst. Mech. Eng. Part B J. Eng. Manuf.* 225, 1051–1062.

Teicher, U., Müller, S., Münzner, J., Nestler, A., 2013. Micro-EDM of carbon fibre-reinforced plastics. *Procedia CIRP* 6, 320–325.

Thao, O., Joshi, S.S., 2008. Analysis of heat affected zone in the micro-electric discharge machining. *Int. J. Manuf. Technol. Manag.* 13, 201. doi: 10.1504/ijmtm.2008.016771.

Yeo, S.H., Kurnia, W., Tan, P.C., Mushan, M., 2007. Development of in situ monitoring and control of micro-EDM process. *Proceedings of the 35th International MATADOR Conference,* Taipei. Springer, 81–84.

9

Modeling of Micro EDM

9.1 Introduction

Understanding and predicting the behavior of a machining process have a vital role in controlling the performance of the process. One of the efficient methods to accomplish that is to model the process with in-depth knowledge of the basic physics behind the material removal mechanism. Compared to conventional machining methods, modeling of the electro discharge machining (EDM) process is complex due to the multiphysics and multi-time scale nature (Maradia and Wegener, 2015). In EDM, the micrometer-scale machining region is comprised of solid, liquid, gas, and plasma states. All the material state may exist together for a short period in the range of microseconds to nanoseconds. The basic mechanism of material removal can only be explained with the help of thermodynamics, plasma physics, magnetohydrodynamics, and fluid dynamics. To incorporate the principles mentioned earlier for modeling the process, various assumptions and simplifications have to be included in the model.

9.2 Comparison of Micro EDM Modeling and Conventional EDM Modeling

In comparison with traditional EDM, the pulse duration in micro EDM is very short (in the order of nanoseconds). Additionally, the energy density of the plasma channel in micro EDM is much higher than the conventional variant (Zahiruddin and Kunieda, 2012). Factors that are neglected in the macroregime, such as surface tension, viscous force, and magnetic pinch force, have more influence in determining the plasma characteristics of micro EDM. The current and voltage in micro EDM cannot be time averaged as in conventional EDM to calculate the pulse power, as they are the function of time in an RC pulse generator (popularly used for micro EDM). The theories that are used to explain the breakdown phenomena and plasma expansion phenomena in EDM found less effective in micro EDM. For example, the theory of bubble mechanism used to explain the dielectric breakdown in EDM cannot be directly applied in micro EDM. As the pulse duration is short, the slow process of bubble expansion fails to make sense of the rapid breakdown phenomena in micro EDM (Dhanik et al., 2005).

Most of the models available for conventional EDM are related to cathode erosion. According to DiBitonto et al. (1989), even though the machining process starts with a rapid erosion of anode surface due to the bombardment of accelerated electrons, the material removal slows down afterward due to the expansion of the plasma channel. The expansion of the plasma radius reduces the local heat flux near the anode surface. Moreover, as the plasma radius is small at or near to the cathode surface, the power density will be higher and creates a deeper crater. So, the workpiece is connected to the cathode during conventional EDM. However, the plasma radius expansion at the anode is comparatively less significant in micro EDM as the discharge duration is very short. The rapid material removal due to the impact of high-velocity electrons will be prominent in this condition. So, the workpiece in micro EDM is connected to the anode. Applying the cathode erosion model in EDM, suggested by DiBitonto et al. (1989) in micro EDM, results in serious prediction errors (Yeo et al., 2007).

Micro EDM plasma generates 30 times higher power density than the conventional EDM plasma (Zahiruddin and Kunieda, 2012). Higher power density results in higher material removal efficiency or plasma flushing efficiency (PFE) and less heat loss to the surrounding fluid medium. In addition, the heat carried away by the debris particles cannot be neglected in the micro EDM. The debris particle reattachment to the tool surface also makes the modeling process more complex. Modeling approaches in micro EDM can be classified as given in Figure 9.1.

Analytical models: These models are developed by analyzing the material removal mechanism using theoretical study. The process is simplified into mathematically interpretable form by selecting appropriate assumptions. The accuracy of the analytical model depends on the chosen assumptions. Analytical models demand an in-depth knowledge of the process. Oversimplification of the process results in a less accurate model, whereas lack of assumptions makes the model more complex and increases the solving time.

Empirical models: In this, extensive experimentation is carried out to correlate the controllable parameters to the machining performance characteristics. During experimentation, the machining variables are changed to analyze the trend in machining output parameters (material removal rate (MRR), tool wear, surface finish, etc.). The input variables are fitted into certain empirical equations that can predict the variations in output parameters. However, the universality of these empirical models is questionable as it largely depends on the machining conditions and machine tool.

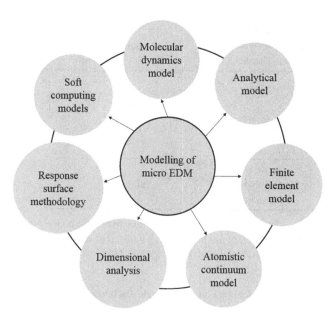

FIGURE 9.1
Modeling approaches in micro EDM.

Numerical models: Numerical methods are used to solve the governing equations of the process with least time and adequate accuracy. The analytical and empirical equations can be used to determine the boundary conditions (BCs). Finite element method (FEM) analysis of the plasma–material interaction using heat conduction equation to generate the thermal distribution profile is popularly used in micro EDM modeling.

9.3 Single-Spark Modeling

Micro EDM modeling can be classified into two types or stages: single-spark modeling and multispark modeling. For calculating the volume of material removal and tool wear in real time, a reliable model has to be developed, which predicts the crater dimensions accurately. The craters are generated by material ablation due to the action of electrical discharge. To calculate the shape and size of the crater produced by individual discharges, the energy transferred to the workpiece or tool has to be known. Figure 9.2 shows the machining zone during the action of a single discharge. The heat from the plasma is dissipated into the anode, cathode, and dielectric fluid in different proportions. F_a stands for a fraction of discharge energy transferred to the anode, F_c stands for the heat transferred to the cathode, and F_d denotes

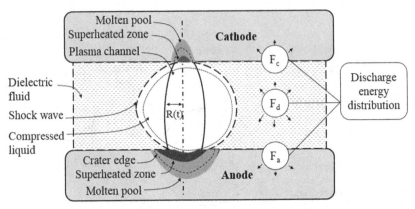

F_c = Fraction of discharge energy transferred to cathode
F_d = Fraction of discharge energy transferred to dielectric
F_a = Fraction of discharge energy transferred to anode

FIGURE 9.2
Schematic diagram of the machining zone in a single discharge micro EDM process.

the energy dissipated to the dielectric fluid. To generate the temperature distribution on the tool and the workpiece, certain assumptions have to be considered, and some BCs have to be assigned. To simulate the heat flux-affected area, the intensity of energy, distribution of energy density, and plasma radius are to be defined. Finally, the molten pool is simulated by generating the temperature profile and recognizing the points where the temperature is greater than the melting temperature/boiling temperature of the specific workpiece material. The overview of a single-spark model to generate crater geometry is shown in Figure 9.3. The modeling process can be divided into two sections: plasma channel modeling and melt pool modeling.

9.3.1 Plasma Channel Modeling

Formation of plasma channel in EDM is mainly explained by the theory of bubble mechanism (Andre, 2011) or electron impact ionization mechanism (Beroual, 1993; Horsten et al., 1971). From the early stages of development of EDM technology, analytical and numerical models are developed to define the nature of EDM plasma. As the process of plasma channel formation is complex, it has been divided into three stages for better understanding: the breakdown phase, heating phase, and material removal phase (Dhanik et al., 2005). In the first stage, the potential barrier between the two electrodes is reduced by various factors, including the presence of positive ions in the interelectrode gap and high density of charged particles on the micropeaks of the cathode. The local electron density is increased as the potential barrier weakens, and bombardment of the electron with the dielectric fluid

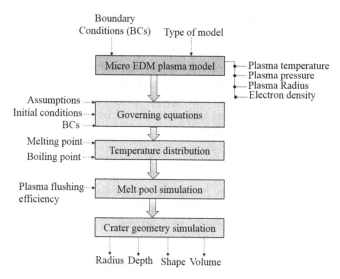

FIGURE 9.3
Overview of the single-spark modeling process.

particles will lead to heating of liquid and the formation of vapor bubbles at the micropeaks of the cathode. The bubble is further expanded, and the bubble pressure reaches a threshold value, leading to a discharge (Watson, 1985). The bubble mechanism model suggests that the local instability due to liquid inhomogeneity and electrode surface irregularity are the main reasons behind the breakdown of the dielectric. According to the electron impact ionization model, the ionization of molecules due to accelerating electrons causes a dielectric breakdown (Beroual, 1993; Horsten et al., 1971).

The bubble mechanism can be effectively used to explain the discharge phenomena, where the discharge time is large. However, for short pulses (in the order of nanoseconds), the bubble expansion theory is insufficient to explain the physics behind rapid discharges. To explain the rapid breakdown phenomena, the theory of gas breakdown (streamer model) can be utilized (Kunhardt and Tzeng, 1988). According to the streamer model, the breakdown happens when the growing electron avalanche transforms into a streamer. The breakdown phenomenon in the extremely small spark gap width is studied by Schoenbach et al. (2008).

To adapt the streamer model to the liquid environment of the dielectric fluid, Dhanik and Joshi (2005) proposed a modified breakdown model. According to this model, the breakdown process begins with nucleation of bubbles at the cathode and expands eventually due to increased density of charged particles. When the bubble characteristics meet the electron impact criteria, instantaneous ionization of the dielectric fluid column occurs, leading to breakdown. The modified model combines the bubble mechanism and streamer model to explain the micro EDM plasma more efficiently

(Chu et al., 2016). The characterization of discharge plasma is done by calculating the electron density and temperature in the plasma column. These
plasma parameters can be calculated using analytical models (Mujumdar
et al., 2013), optical spectroscopy (Nagahanumaiah et al., 2009), photomulti-
pliers (Albinski et al., 1996), etc. Furthermore, understanding the mechanism
behind plasma formation helps to explain some unique phenomena related
to EDM, including the cooling effect produced by elevated iron content in
the plasma (Adineh et al., 2012).

9.3.1.1 Modeling of Conventional EDM Plasma

Generally, the plasma channel modeling can be done in two different ways:
the fluid dynamic approach and kinetic approach. In the fluid dynamic
approach, the plasma parameters are determined by solving fluid dynamic
and heat transfer equations. In the kinetics approach, the density and
distribution of various particles in the plasma are calculated to analyze the
interaction of those particles with the cathode/anode surface.

In the fluid dynamics model (Dhanik et al., 2005), certain assumptions are
incorporated for simplification as follows:

1. Uniform temperature, pressure, and density throughout the plasma
   channel
2. Time-independent properties of dielectric fluid
3. Cylindrical-/spherical-shaped plasma
4. The role of particle bombardment in heat transfer is neglected
5. Uniform expansion of plasma
6. Energy fraction to the plasma is constant.

Early attempts in the modeling of plasma channel in conventional EDM
include Van Dijk and Snoeys (1971)'s cylindrical plasma model, where a constant mass cylindrical plasma approximation is used to define the discharge
process. Electron emission theory is used to calculate the power distribution
during plasma generation. Continuity and momentum equations are used
to determine the plasma parameters. This model suggested superheating
as the material removal mechanism in EDM. However, the model fails to
accommodate the effects of radiative heat transfer as well as heat loss during
dielectric fluid vaporization and ionization. As a result, the model underes-
timates the plasma pressure. Lhiaubet and Meyer (1981) introduced a new
model in which the expanding spherical plasma is used. In this model, the
properties of diatomic plasma are kept temperature independent. This cathode erosion model did not consider the pressure variation parameter in the
energy balance equations. During experimental validation, the temperature
and pressure of the plasma are found to be overestimated. The model developed by Eubank et al. (1993) introduced a variable mass cylindrical plasma.

Time-dependent plasma radius, temperature, and pressure are estimated. One of the main challenges in modeling the EDM plasma is the difficulty in incorporating the unsteady behavior of plasma expansion (due to variation in macroscopic properties of the plasma and interionic bombardments). Eubank et al. (1993) used an unsteady energy balance equation with varying mass to incorporate the unsteady nature of EDM plasma. Solving the continuity equation and momentum equation resulted in the formulation of a relationship connecting plasma radius and time. In the fluid dynamics method, heat transfer from plasma is considered as the sole reason for material removal. The kinetics model deals with the analysis of density and distribution of different particles in the plasma. Assuming that the ions and electrons follow Boltzmann's distribution, analyzing the electron generation using Poisson's equation and solving the mass and momentum equations will lead to the prediction of the volume of material removal due to electromechanical erosion.

To simulate the crater profile on the workpiece and tool, certain plasma BCs have to be well defined. This includes

- Plasma radius
- Heat source shape
- Heat intensity distribution profile
- Plasma temperature distribution
- Plasma pressure
- Electron density.

Modeling of the plasma channel is carried out to identify the BCs for the crater simulation. The BCs are also established by experimental methods including optical spectroscopy. EDM models use high-voltage, high-current discharges with long discharge time to estimate the plasma radius, which often results in overestimation. Eubank et al. (1993) formulated the relationship between the radius of the cylindrical plasma and discharge duration from their plasma model for low-current, low-voltage discharges in the small spark gap. Equation (9.1) shows the plasma radius suggested by Eubank et al. (1993):

$$R = 0.788 t_{on}^{3/4} \qquad (9.1)$$

where R is the plasma radius and t is the discharge period.

Pandey and Jilani (1986) assumed that the cathode spot temperature is equal to the boiling point of the material (T_b) and derived Eq. (9.2) connecting the spot temperature and plasma radius (R).

$$T_b = \frac{E \times F_c \times 10^6}{k \times R \times \pi^{3/2}} \tan^{-1} \left[\frac{4 \alpha t_{on} \times 10^6}{R^2} \right]^{\frac{1}{2}} \qquad (9.2)$$

where T_b is the boiling point temperature in kelvin, R is the heat source radius in meters, k is the thermal conductivity in watts per millikelvin, α is the thermal diffusivity in square meter per second, t is the pulse duration in microseconds, E is the total discharge power in watts, and F_c is the cathode energy fraction.

To derive the temperature distribution at a point in the plasma, the heat flux in the plasma column has to be known. Various researchers used different heat flux assumptions such as constant heat flux and Gaussian heat flux. The direct plasma analysis (high-speed imaging and spectroscopy) technique revealed that the maximum heat intensity is at the center of the plasma, which makes the Gaussian heat flux more suitable for modeling the heat source.

9.3.1.2 Modeling of Micro EDM Plasma

One of the early approaches to model the micro EDM plasma channel and interaction of plasma with electrode material is done by Katz and Tibbles (2005). In this model, the input variables including discharge current, discharge voltage, spark duration, dielectric permittivity, and electrode diameter are formulated as a dimensionless group. Least-square regression method is used to calculate electrode current density, microcrater area, channel power dissipation, and rate of expansion of plasma channel. The pioneer work in modeling the micro EDM plasma by considering different stages of discharge formation is carried out by Dhanik and Joshi (2005).

As discussed earlier, the theory of bubble mechanism failed to cover the complete aspects of the plasma generation process in micro EDM. The velocity of plasma channel propagation is found to be in the order of 10^5 m/s, which cannot be described by the slow hydrodynamic expansion vapor bubble. So, Dhanik and Joshi (2005) suggested a combination of bubble mechanism (bubble initiation) and electron impact ionization (bubble expansion and dielectric breakdown) mechanism to explain the rapid breakdown phenomena in micro EDM. The BCs for plasma modeling in micro EDM is shown in Figure 9.4 (Chu et al., 2016). Inside the spark gap, charged particles are accelerated towards opposite polarity followed by an increase in plasma radius. However, the expansion is restricted by body forces at the plasma boundary.

General steps in the modeling of micro EDM plasma by Dhanik and Joshi (2005) towards finding the heat dissipated on the electrode surface is shown in Figure 9.5.

Experimental investigation of the plasma temperature and electron density is also done with the help of optical emission spectroscopy (Subbu et al., 2012). The plasma model suggested by Dhanik and Joshi (2005) used a cylindrical plasma in water as a dielectric medium, similar to Eubank et al. (1993). Mujumdar et al. (2013) pointed out that the model lacks a deep understanding of the plasma chemistry in terms of identifying the various species in the plasma and their effect on the plasma characteristics. The usage of

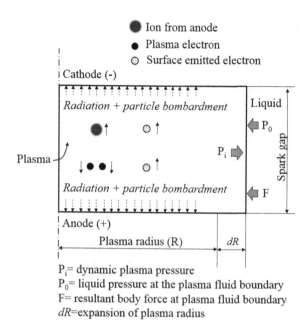

FIGURE 9.4
BCs in plasma channel modeling.

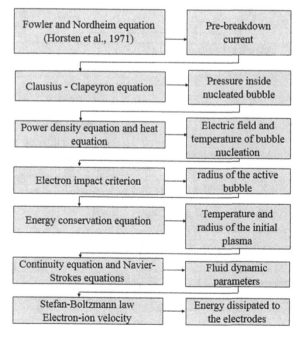

FIGURE 9.5
Steps of modeling EDM plasma channel.

cylindrical plasma instead of spherical plasma in the previous models is also questioned by Mujumdar et al. (2013).

A global plasma model approach by assuming a single-pulse micro EDM discharge between parallel plates is realized (Mujumdar et al., 2013). In this model, the densities of the species in the plasma are volume averaged, and the Maxwellian energy distribution is applied. By coupling the particle balance equation, plasma dynamics, and energy balance, the plasma temperature and pressure are calculated. The average electron temperature predicted by the model is around 6,817 K, and the electron density is $8.9 \times 10^{23}\,\mathrm{m^{-3}}$.

Even though these models could incorporate the physics of the micro EDM plasma, they failed to represent the actual phenomena by neglecting certain aspects of the process. According to Chu et al. (2016), most of the researchers omitted the role of magnetic pinch force, viscous force, and surface tension of the fluid in their model. Chu et al. (2016) proposed a comprehensive model to predict the plasma radius, pressure, and temperature, incorporating the factors mentioned earlier. Steps in the modified model of plasma channel developed by Chu et al. (2016) are shown in Figure 9.6. Plasma radius, plasma temperature, and plasma pressure are used to model the heat source and to provide the BCs in the modeling of the crater surface.

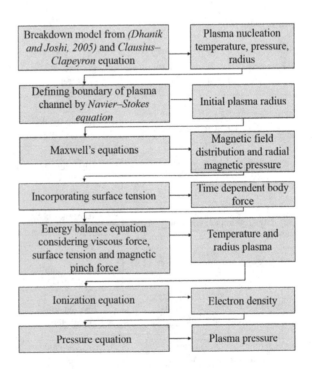

FIGURE 9.6
Steps of modeling plasma radius, pressure, and temperature considering surface tension, viscous force, and magnetic pinch force.

9.3.1.3 Plasma Parameters (Energy Fraction, Heat Flux, Spark Radius)

To simulate the erosion of material due to interaction with plasma, the BCs have to be defined accurately. Plasma radius, heat flux distribution, energy partition factor, discharge intensity, electron density, plasma pressure, and plasma temperature are defined in the BCs for solving the governing equations in the model.

Energy partition factor: The discharge energy is partitioned between the electrodes, dielectric fluid, and debris particles as shown in Figure 9.7. Determining the fraction of energy transferred to the electrode plays a vital role in calculating the temperature distribution in the electrodes. The energy distribution ratio depends on the electrode materials, dielectric fluid, discharge energy, and fluid pressure (Hoang et al., 2015). The energy carried away by the debris particles is often neglected in the case of macro EDM as significant heat dissipation occurs in the electrodes and dielectric fluid. However, in micro EDM, the short-pulse duration results in high power density and high material removal efficiency. So, a considerable part of the heat is also dissipated to the debris particles.

Zahiruddin and Kunieda (2010) calculated the energy distribution ratio in micro EDM, incorporating the heat loss to the debris particles. According to the study, the percentage of energy distribution during micro EDM to Cu foil anode is in the range of 6.6%–10.8%, whereas cathode gets 3.3%–3.7%. With the help of inverse analysis correlating an electrothermal model and crater measurements from single-spark experiments, Shao and Rajurkar (2013) estimated the energy partition to the anode as 9.4% and to the cathode as 3.6%. Furthermore, a relationship between the energy fraction at anode, cathode, and discharge parameter is developed by regression analysis (Hoang et al., 2015). An energy fraction of 10% is found to be optimum for Ti anode and 8% for Ti cathode. Bigot et al. (2016) analyzed the energy fraction in micro EDM by dividing the total energy consumed by the electrodes

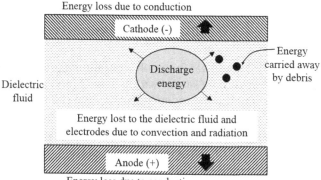

FIGURE 9.7
Partition of the discharge energy in the spark gap.

TABLE 9.1

Energy Fraction Used by Different Researchers to Model Plasma as a Heat Source

Authors	EDM/μ-EDM	Anode (%)	Cathode (%)
Van Dijck (1973)	EDM	50	50
Pandey and Jilani (1986)	EDM	50	50
DiBitonto et al. (1989)	EDM	8	18.3
Patel et al. (1989)	EDM	8	18
Xia et al. (1996)	EDM	40–48	25–34
Revaz et al. (2005)	EDM	–	10–15
Zahiruddin and Kunieda (2010)	μ-EDM	6.6–10	3.3–3.7
Singh (2012)	EDM	–	6–27
Zhang et al. (2013)	EDM	–	44
Shabgard et al. (2013)	EDM	4–36	4–9
Shao and Rajurkar (2013)	μ-EDM	9.4	3.6
Hoang et al. (2015)	μ-EDM	10	8
Bigot et al. (2016)	μ-EDM	11.09	1.19
Maradia et al. (2015)	EDM	28–36	15–45

into the energy required to elevate the electrode temperature to the melting point, latent heat for melting, energy for attaining vaporization temperature, and latent heat of vaporization. According to the study, heat dissipated to the anode is 11.09% and to the cathode is 1.19%. Different energy fraction values used by researchers in EDM and micro EDM are summarized in Table 9.1.

Heat flux: Heat flux is defined as a spatial distribution of heat intensity in the plasma channel. The type of heat flux also makes a significant difference in the simulation results. Researchers defined plasma as a heat source with different heat distributions.

Commonly, following types of heat flux have been used for modeling of the heat source.

1. Point heat source: In this, the heat load is assumed to be applied at a single point.
2. Uniformly distributed heat flux with constant magnitude: Heat flux density is assumed to be same over the area. Moreover, the energy intensity is assumed to be independent of time (Jilani and Pandey, 1982).
3. Uniformly distributed heat flux with time-dependent magnitude: The heat flux is distributed evenly on the plasma column, but the quantity varies with time (Patel et al., 1989).
4. Gaussian heat flux with constant magnitude: Here, the heat flux density is assumed to be maximum at the center and minimum at the periphery of the plasma channel. The variation of heat flux follows

a Gaussian function. However, the magnitude of energy density at a point is assumed to be constant during discharge (Das et al., 2003).

5. Gaussian heat flux with time-varying magnitude: The magnitude of energy density at a point is assumed to vary over time. This is considered as a more realistic approach to model the heat source, as the plasma is known to be expanding during the discharge period (Tan and Yeo, 2008).

From the experimental investigation of the discharge plasma with high-speed imaging and light emission spectroscopy, it is observed that light intensity and electron density are maximum at the center of the plasma channel. So, Gaussian heat flux is the most realistic assumption to be used for modeling. However, solving the equations with Gaussian heat flux conditions demands more computational resources. Therefore, some researchers used simplified heat flux approximations. An overview of the heat flux used in EDM and micro EDM modeling is provided in Table 9.2.

Once the heat flux is selected, the spark radius is estimated from plasma models and used as an input to calculate the magnitude of the heat flux at

TABLE 9.2

Heat Flux Used in Different Models in EDM/Micro EDM

Authors	Heat Flux Type	Equation, q (W/m²) =
Snoyes and Van Dijck (1971), Beck (1981), Pandey and Jilani (1986), and Patel et al. (1989)	Disc-type heat flux (EDM)	$\dfrac{F_c VI}{\pi r_c^2}$
Zhang et al. (2013), Shabgard et al. (2013), Allen and Chen (2007), and Kansal et al. (2008)	Gaussian heat flux (EDM)	$\dfrac{4.57 FVI}{\pi R(t)^2} \exp\left(-4.5\left(\dfrac{r}{R(t)}\right)^2\right)$
Shao and Rajurkar (2015a, b), Murali and Yeo (2005), Dilip et al. (2019), Somashekhar et al. (2013), and Kuriachen et al. (2015)	Gaussian heat flux (μ-EDM)	$\dfrac{3.1572 FVI}{\pi R(t)^2} \exp\left(-3\left(\dfrac{r}{R(t)}\right)^2\right)$
Mujumdar et al. (2014)	Gaussian heat flux (μ-EDM)	$\dfrac{q_0(t) R(t)^2}{2\sigma_R^2\left[1-\exp\left(\dfrac{-R(t)^2}{\sigma_R^2}\right)\right]} \exp\left(-\left(\dfrac{r}{2\sigma_R}\right)^2\right)$
		$\sigma_R = \dfrac{\text{Radius of cathode}}{3}$
Kumar and Yadava (2008)	Gaussian heat flux (μ-EDM)	$\dfrac{FVI}{\pi R(t)^2} \exp\left(-2\left(\dfrac{r}{R(t)}\right)^2\right)$
Tan and Yeo (2008)	Uniform heat flux (μ-EDM)	$\dfrac{FVI}{\pi R(t)^2}$

F is the fraction of energy transferred to respective electrodes, $R(t)$ is the time-dependent plasma radius, r is the radial distance of a point from the origin, rc is the radius of plasma at cathode, $q_0(t)$ is the maximum heat flux at $r = 0$, I is discharge current, and V is the discharge voltage.

different locations. Spark radius can also be found out using empirical relationships formulated via extensive experimentation. However, selecting the
empirical formula for the plasma radius is challenging as the spark radius
is affected by the type of electrodes, dielectric medium, and flow pressure.
Error in the estimation of spark radius will affect the accuracy of heat flux
magnitude and eventually reflected in the accuracy of the predicted crater
dimensions. Spark radius equations used by different researchers are consolidated in Table 9.3.

Plasma flushing efficiency: The ratio of the actual volume removed from
the crater to volume of the molten pool is known as PFE. The boundary of
the crater is often defined at the isothermal line above the melting temperature of the material. However, all the melted material may not be flushed out
from the machining zone, which brings significant errors in the calculation
of crater dimensions. To realize an accurate model, PFE should be incorporated into the model.

TABLE 9.3

Equations Used in Different EDM/Micro EDM Models to Estimate Plasma Radius

Authors	Plasma Radius $R(t)$ (µm) =	EDM/µ-EDM	Remarks
Shabgard et al. (2013)	$2.04e^{-3}\left(I^{0.43}t_{on}^{0.44}\right)$	EDM	The empirical equation is derived by Ikai and Hashigushi (1995)
Somashekhar et al. (2013)	Constant spark radius	µ-EDM	Multispark model
Kansal et al. (2008)	$ZP^m t_{on}^n$	µ-EDM	Developed by solving the mathematical model of DiBitonto et al. (1989) by integration (Erden, 1983)
Tan and Yeo (2008)	$0.059t_{on}^{0.79}$	µ-EDM	Empirical model developed from single-spark experiments
Kumar and Yadava (2008)	–	µ-EDM	Plasma channel modeling by Dhanik and Joshi (2005) is used to calculate the plasma radius
Allen and Chen (2007)	Constant spark radius (5 µm)	µ-EDM	–
Dilip et al. (2019)	Constant spark radius (20% of the tool radius)	µ-EDM	By analyzing the literature (Singh, 2012; Tao et al., 2012)
Yeo et al. (2007) and Zhang et al. (2015)	$R_{anode} = 0.0284t_{on}^{0.9115}$ $R_{cathode} = 0.0425t_{on}^{0.0895}$	µ-EDM	Empirical equation
Shao and Rajurkar (2013)	$0.788t_{on}^{0.75}$	µ-EDM	Empirical equation
Hoang et al. (2015)	$R_0\rho^a I^b t_{on}^c$	µ-EDM	R_0, a, b, and c can be found out empirically by analysis of machined crater

I is the discharge current, *P* is the discharge power, t_{on} is the discharge period, ρ is the dielectric
material density, and Z, *m*, *n*, R_0, *a*, *b* and *c* are the empirical constants.

9.4 Simulation of Crater Formation

As explained earlier, the material removal in micro EDM can be explained with the electrothermal model or electromechanical model. In the electrothermal model, the melting and superheating of the electrode material are responsible for material removal. Solving the partial differential equations related to heat transfer will provide the temperature distribution on the surface. On the other hand, analysis of the electric field strength in the discharge plasma helps to calculate the surface stress due to particle bombardment. Material rupture on the surface due to the impact of charged particles can be simulated to get the dimensions of the discharge crater. Most of the attempts to model the crater formation are based upon electrothermal models.

9.4.1 Assumptions

Major assumptions in electrothermal modeling of micro EDM are given as follows:

1. Homogeneity of the tool and workpiece material.
2. The material properties of the tool and workpiece can be temperature dependent or independent.
3. For simplification, some researchers assumed single plasma channel formation per pulse.
4. Energy loss due to radiation is neglected in most of the models.
5. The assumptions are related to heat flux (shape (circular/pointed), type (uniform/Gaussian), plasma radius (constant/time dependent)).
6. The assumptions are related to heat transfer modes (convection and radiation to the dielectric, conduction to electrodes).
7. PFE (ratio of volume removed from crater to the volume of melt pool) is to be mentioned.
8. While conducting a 2D analysis of the plasma–electrode interaction, half of the interaction zone is modeled, and the axis-symmetrical assumption is mentioned.

9.4.2 Governing Equations

Heat transfer: Fourier's law of conductive heat transfer equation, Eq. (9.3), can be used to find out the temperature distribution in the discharge zone.

$$\frac{\partial^2 T}{\partial x^2} + \frac{\partial^2 T}{\partial y^2} + \frac{\partial^2 T}{\partial z^2} + \frac{\dot{q}}{k} = \frac{1}{\alpha}\frac{\partial T}{\partial t} \tag{9.3}$$

where T is the temperature, x-y-z are the Cartesian coordinates, $\dot{q}$ is the internal energy generation per unit volume (Joule effect), k is the coefficient of thermal conductivity, α is the thermal diffusivity, and t stands for time. Thermal diffusivity α is given by Eq. (9.4):

$$\alpha = \frac{k}{\rho c_p} \tag{9.4}$$

where ρ is the density and c_p is the specific heat constant.

For solving the heat transfer model numerically, the heat equation is written as in Eq. (9.5):

$$\rho c_p \left(\frac{\partial T}{\partial t} \right) + \rho c_p \bar{u} \cdot \nabla T = \nabla(k\nabla T) + Q \tag{9.5}$$

The BCs include inward heat flow condition, Eq. (9.6),

$$\bar{n} \cdot (-k\nabla T) = q \tag{9.6}$$

All other boundaries are considered as adiabatic walls defined by Eq. (9.7):

$$\bar{n} \cdot (-k\nabla T) = 0 \tag{9.7}$$

where $\bar{u}$ is the velocity vector, n is the normal vector in the outward direction, Q stands for the thermal energy per unit volume, and q is the heat flux per unit area.

When both energy transfer and fluid motion are involved, Navier–Stokes equations based on conservation of mass, momentum, and energy are used to define the physics of material removal (Dilip et al., 2019). Using Navier–Stokes equations, the heat transfer to the workpiece is given by Eq. (9.8) as

$$\frac{\partial}{\partial t}(\rho h) + \nabla \cdot (\rho V h) = \nabla \cdot (k\nabla h) + S_h \tag{9.8}$$

where h is the specific enthalpy, V is the total velocity vector of the flow, and S_h is the factor to include latent heat during melting and vaporization.

9.4.3 Thermal Boundary Conditions on the Electrode Surface

The BCs in the electrothermal model may change according to the heat flux applied. Here, BCs for an electrothermal model with Gaussian heat flux are explained. As shown in Figure 9.8, at boundary 1, a Gaussian heat flux is applied for a length equal to the plasma radius (R) beyond which convective heat transfer occurs from the electrode surface to the dielectric liquid. The BCs at boundary 1 can be summarized mathematically by Eq. (9.9):

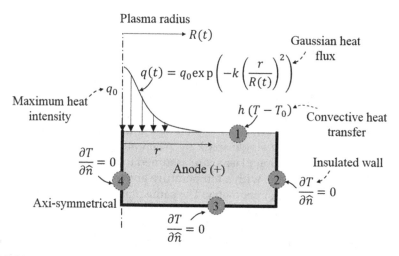

FIGURE 9.8
Thermal BCs in an electrothermal model of micro EDM.

$$
k\frac{\partial T}{\partial y} = \begin{cases} q(t) & r \leq R \\ h(T - T_0) & r \geq R \end{cases} \tag{9.9}
$$

where k is the thermal conductivity, $q(t)$ is the varying heat flux, h is the convective heat transfer coefficient, and T_0 is the ambient temperature. At boundaries 2, 3, and 4, the heat transfer is assumed to be zero as the wall is simplified as an insulated surface, Eq. (9.10).

$$
k\frac{\partial T}{\partial \hat{n}} = 0 \tag{9.10}
$$

where $\hat{n}$ is the normal unit vector.

When the heat source disappears after the discharge period, boundary 1 is wholly subjected to convective heat transfer.

9.4.4 Generation of Temperature Profile and Crater Geometry

To solve the governing equations, different approaches were suggested and tested by researchers. Yeo et al. (2007) solved the heat transfer equation using integral transform method, as given in Eq. (9.11).

$$
T(r,z,t) = \frac{qR}{2K_t} \int_0^\infty J_0(\lambda r) J_1(\lambda r) \times \left[e^{-(\lambda z)} erfc\left(\frac{z}{2\sqrt{\alpha t_{on}}} - \lambda\sqrt{\alpha t_{on}} \right) \right.
$$

$$
\left. -e^{-(\lambda z)} erfc\left(\frac{z}{2\sqrt{\alpha t_{on}}} + \lambda\sqrt{\alpha t_{on}} \right) \right] \frac{d\lambda}{\lambda} \tag{9.11}
$$

where R is the heat flux radius, J_0 and J_1 are Bessel functions, and *erfc* is the complementary error function.

The temperature distribution in the radial and vertical axis ($T(r,z,t)$) is calculated by solving the integral equation. The crater edge temperature directly measured during single-spark experiments is correlated with the temperature profile to approximate the diameter and depth of the crater. The average error in predicting the diameter, depth, and volume was 3.6%, 0.4%, and 6.6%, respectively. ANSYS platform is used to simulate the temperature profile during a single spark by Allen and Chen (2007). Constant heat flux and nonexpanding plasma are assumed to be responsible for material ablation. The elements with a temperature above the melting points are "killed" to create the crater geometry. The same platform is used for predicting the temperature profile and residual stress in titanium alloys by Murali and Yeo (2005). Temperature-dependent material properties are used in the model. Gaussian distribution is used as the heat flux, and heat transfer equations are solved to find out the diameter-to-depth ratio of the microcraters. Finite element approximation equations to solve the heat transfer model for micro EDM are suggested by Kumar and Yadava (2008). They have studied the effect of spark radius on crater dimensions and MRR. A modified melt pool model for micro EDM is developed by Mujumdar et al. (2014). A level set method is utilized to distinguish the solid and liquid fractions in the workpiece when melted by a Gaussian heat flux. The heat transfer model is solved with a COMSOL Multiphysics software®. The stress in the surface due to surface tension gradient (Marangoni stress) is also included to define the fluid motion in the melt pool.

Combining the heat transfer equations and fluid flow equations, the crater geometry is simulated by Shao and Rajurkar (2015b). The role of surface tension gradient in determining the crater geometry is also emphasized in their model. A two-dimensional electrothermal model is solved using a finite volume method by Kuriachen et al. (2015). The transient thermophysical model is used to plot 3D temperature distribution profiles. According to Dilip et al. (2019), the reason for the irregular profile of the crater is the motion of the top and subsequent layers of the molten pool. The fluid motion in the mushy zone (where the solid and liquid state coexists) in the discharge crater is incorporated in their model for predicting the crater geometry accurately. Validation of the model shows that the error in prediction of crater diameter is in the range of 5.19%–5.64% and error in crater depth is in the range of 10.58%–17.47%. Figure 9.9 shows the simulation of a crater pool during single-spark micro EDM (Dilip et al., 2019).

The material removal process of micro EDM is also analyzed with a molecular dynamics model by Yang et al. (2011). The PFE is identified as 0.02–0.05, which means a large amount of molten pool is resolidified during crater formation. The bulge formation during machining is explained by ejection of molten material from the crater center due to high-pressure plasma. Combining the molecular dynamics model with temperature models, an

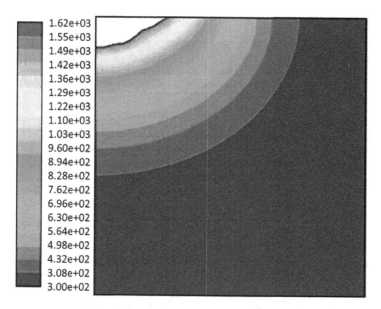

FIGURE 9.9
Temperature profile and crater geometry simulated for single discharge (Dilip et al., 2019).

atomic continuum model is established by Guo et al. (2014). The material removal phenomena in micro EDM is thoroughly analyzed by studying the temperature, stress, and material microstructure evolution during the action of thermal load and particle bombardment.

9.5 Multispark Modeling

Once the dimensions of the crater are found by solving the governing equations analytically or numerically, it can be used to calculate the topography of the surface machined by micro EDM. For that, multiple sparks have to be generated on the workpiece surface. Figure 9.10 shows the general steps in the multispark modeling of micro EDM process. Kurnia et al. (2009) attempted to model the surface roughness in micro EDM with the help of nonoverlapping multiple sparks. The craters are assumed to be densely packed in line without any bulging on the rim. The reattachment of debris is neglected along with the effect of microcracks. The surface roughness values are calculated from mathematical analysis of the geometry of the craters aligned without overlap. Another multispark model developed by Kiran and Joshi (2007) discusses the effect of debris particles on the surface roughness. The micropeaks in the workpiece surface are assumed to be normally distributed, and

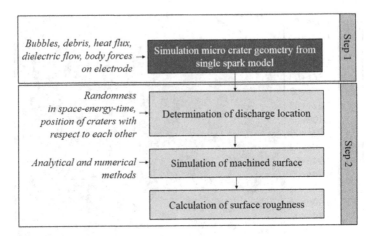

FIGURE 9.10
Steps in the multispark model for simulation of machined surface topography.

the spark impinges on random points with a minimum distance between the tool and the workpiece. The distribution of debris particles in the dielectric medium is incorporated into the model. As the presence of debris particles affects the electrode potential and plasma composition, the radius of the melt pool for a single spark is changed accordingly. These calculations are used in the multispark model to alter the surface roughness model for getting more realistic results.

An FEM model for overlapping craters is developed by Tan and Yeo (2008). The machined surface topography is extracted based on maximum asperity condition (highest surface asperity with a pointed top-deepest crater profile) and minimum asperity condition (shallowest crater profile). The distance separating two craters is assumed to be equal to the heat source diameter, and the 2D model is solved for calculating the surface roughness. A finite volume method on a uniform grid with implicit flux discretization is used to solve the multispark model by Somashekhar et al. (2013). However, most of the multispark model for micro EDM is oversimplified and fails to give the real picture of the random multisparks. Jithin et al. (2017) developed a more sophisticated 3D multispark model for conventional EDM, where the 3D surface roughness is calculated by solving a single discharge model. The stochastic nature of the EDM process is well represented by including randomness of space, energy, and time of the discharges. PFE is also considered for the crater geometry calculations. Different interactions of the simultaneous crater are simulated, including craters with/without overlap and craters touching boundaries. Figure 9.11 shows a machined surface simulated by multispark modeling in EDM (Izquierdo et al., 2009). However, no attempts are yet reported to model 3D surface topography in micro EDM'ed surface, incorporating the randomness in space, energy, and time.

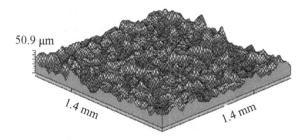

FIGURE 9.11
Surface simulated by the multispark model in EDM (Izquierdo et al., 2009).

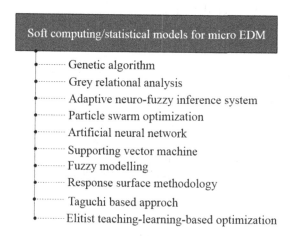

FIGURE 9.12
Soft computing/statistical model approaches in micro EDM.

9.6 Soft Computing Models

To model the micro EDM process, several soft computing techniques are utilized by different researchers. Most of the research works are focused on predicting the performance parameters and optimizing the machining conditions for better machining characteristics. An overview of the soft computing techniques used in micro EDM is provided in Figure 9.12.

9.7 Other Models

Apart from the modeling and simulation of crater geometry and surface topography, some other models are also available related to micro EDM.

Change in the feature geometry in reverse micro EDM due to the presence of debris particle and secondary discharges is simulated using numerical modeling by Roy et al. (2018). The single-spark model is dynamically modified by considering the reattachment of debris particles on the tool surface. Modeling of the vibration-assisted micro EDM process is carried out by Jahan et al. (2010). Model of debris motion is realized by Joshi et al. (2014) in vibration-assisted micro EDM and Mastud et al. (2015) in reverse micro EDM. Simulation of the micro EDM deposition process is developed by Wang et al. (2011) using an electrothermal model. A model-based analysis of the energy consumption in micro EDM drilling is carried out by Franco et al. (2016). A mathematical model to predict manufacturing cost in micro EDM drilling is developed by D'Urso et al. (2017).

9.8 Conclusion

Modeling of micro EDM process helps to create a deep understanding of the machining behavior. The primary step is to model the plasma channel to get the plasma radius, temperature, and pressure. These parameters are used to calculate the energy dissipated to the electrode surface. Using the theory of heat transfer, the temperature profile can be simulated afterward. Finally, the crater geometry is simulated by analyzing the areas above the melting temperature of the electrode material. Assumptions incorporated in the model can decide the accuracy of the calculation of crater dimensions. To increase the accuracy of the model, fluid dynamic equations can be incorporated. Apart from thermal modeling, researchers also attempted to model the process using electromechanical models, where the theory of electron–ion impact mechanism is used. Calculated crater dimensions can be used to simulate the machined surface using multispark modeling. In addition to the theoretical and numerical modeling, soft computing models are also developed to correlate the machining parameters and performance parameters.

References

Adineh, V., Coufal, O., Zivny, O., 2012. Thermodynamic and radiative properties of plasma excited in EDM process through N2 taking into account Fe. *IEEE Trans. Plasma Sci.* 40, 2723–2735.

Albinski, K., Musiol, K., Miernikiewicz, A., Labuz, S., Malota, M., 1996. The temperature of a plasma used in electrical discharge machining. *Plasma Sources Sci. Technol.* 5, 736.

Allen, P., Chen, X., 2007. Process simulation of micro electro-discharge machining on molybdenum. *J. Mater. Process. Technol.* 186, 346–355.

Andre, D., 2011. Conduction and breakdown initiation in dielectric liquids. *2011 IEEE International Conference on Dielectric Liquids*, Trondheim, Norway. IEEE, 1–11.

Beck, J.V., 1981. Large time solutions for temperatures in a semi-infinite body with a disk heat source. *Int. J. Heat Mass Transf.* 24, 155–164.

Beroual, A., 1993. Electronic and gaseous processes in the prebreakdown phenomena of dielectric liquids. *J. Appl. Phys.* 73, 4528–4533.

Bigot, S., D'Urso, G., Pernot, J.P., Merla, C., Surleraux, A., 2016. Estimating the energy repartition in micro electrical discharge machining. *Precis. Eng.* 43, 479–485.

Chu, X., Zhu, K., Wang, C., Hu, Z., Zhang, Y., 2016. A study on plasma channel expansion in micro-EDM. *Mater. Manuf. Process.* 31, 381–390.

D'Urso, G., Quarto, M., Ravasio, C., 2017. A model to predict manufacturing cost for micro-EDM drilling. *Int. J. Adv. Manuf. Technol.* 91, 2843–2853.

Das, S., Klotz, M., Klocke, F., 2003. EDM simulation: Finite element-based calculation of deformation, microstructure and residual stresses. *J. Mater. Process. Technol.* 142, 434–451.

Dhanik, S., Joshi, S.S., 2005. Modeling of a single resistance capacitance pulse discharge in micro-electro discharge machining. *J. Manuf. Sci. Eng.* 127, 759.

Dhanik, S., Joshi, S.S., Ramakrishnan, N., Apte, P.R., 2005. Evolution of EDM process modelling and development towards modelling of the micro-EDM process. *Int. J. Manuf. Technol. Manag.* 7, 157.

DiBitonto, D.D., Eubank, P.T., Patel, M.R., Barrufet, M.A., 1989. Theoretical models of the electrical discharge machining process. I. A simple cathode erosion model. *J. Appl. Phys.* 66, 4095–4103.

Dilip, D.G., Ananthan, S.P., Panda, S., Mathew, J., 2019. Numerical simulation of the influence of fluid motion in mushy zone during micro-EDM on the crater surface profile of Inconel 718 alloy. *J. Braz. Soc. Mech. Sci. Eng.* 41, 1–14.

Erden, A., 1983. Effect of materials on the mechanism of electric discharge machining (EDM). *J. Eng. Mater. Technol.* 105, 132–138.

Eubank, P.T., Patel, M.R., Barrufet, M.A., Bozkurt, B., 1993. Theoretical models of the electrical discharge machining process. III. The variable mass, cylindrical plasma model. *J. Appl. Phys.* 73, 7900–7909.

Franco, A., Rashed, C.A.A., Romoli, L., 2016. Analysis of energy consumption in micro-drilling processes. *J. Clean. Prod.* 137, 1260–1269.

Guo, J., Zhang, G., Huang, Y., Ming, W., Liu, M., Huang, H., 2014. Investigation of the removing process of cathode material in micro-EDM using an atomistic-continuum model. *Appl. Surf. Sci.* 315, 323–336.

Hoang, K.T., Gopalan, S.K., Yang, S.H., 2015. Study of energy distribution to electrodes in a micro-EDM process by utilizing the electro-thermal model of single discharges. *J. Mech. Sci. Technol.* 29, 349–356.

Horsten, H.J.A., Heuvelman, C.J., Veenstra, P., 1971. An introductory investigation of the breakdown mechanism in electro discharge machining. *Ann. CIRP* 20, 43–44.

Ikai, T., Hashigushi, K., 1995. Heat input for crater formation in EDM. *Proceedings of the International Symposium for Electro-Machining-ISEM XI*, Lausanne, Switzerland. EPFL, 163–170.

Izquierdo, B., Sanchez, J.A., Plaza, S., Pombo, I., Ortega, N., 2009. A numerical model of the EDM process considering the effect of multiple discharges. *Int. J. Mach. Tools Manuf.* 49, 220–229.

Jahan, M.P., Saleh, T., Rahman, M., Wong, Y.S., 2010. Development, modeling, and experimental investigation of low frequency workpiece vibration-assisted micro-EDM of tungsten carbide. *J. Manuf. Sci. Eng.* 132, 054503.

Jithin, S., Bhandarkar, U. V., Joshi, S.S., 2017. Analytical simulation of random textures generated in electrical discharge texturing. *J. Manuf. Sci. Eng.* 139, 111002.

Jilani, S.T., Pandey, P.C., 1982. Analysis and modelling of EDM parameters. *Precis. Eng.* 4, 215–221.

Joshi, S.S., Singh, R.K., Samuel, J., Kothari, N.S., Mastud, S.A., 2014. Analysis of debris motion in vibration assisted reverse micro electrical discharge machining. *Mater. Micro Nano Technol. Prop. Appl. Syst. Sustainable Manuf.* 24, V001T03A017.

Kansal, H.K., Singh, S., Kumar, P., 2008. Numerical simulation of powder mixed electric discharge machining (PMEDM) using finite element method. *Math. Comput. Model.* 47, 1217–1237.

Katz, Z., Tibbles, C.J., 2005. Analysis of micro-scale EDM process. *Int. J. Adv. Manuf. Technol.* 25, 923–928.

Kiran K, Joshi, S.S., 2007. Modeling of surface roughness and the role of debris in micro-EDM. *J. Manuf. Sci. Eng.* 129, 265.

Kumar, R., Yadava, V., 2008. Finite element thermal analysis of micro-EDM. *Int. J. Nanopart.* 1, 224–240.

Kunhardt, E.E., Tzeng, Y., 1988. Development of an electron avalanche and its transition into streamers. *Phys. Rev. A* 38, 1410.

Kuriachen, B., Varghese, A., Somashekhar, K.P., Panda, S., Mathew, J., 2015. Three-dimensional numerical simulation of microelectric discharge machining of Ti-6Al-4V. *Int. J. Adv. Manuf. Technol.* 79, 147–160.

Kurnia, W., Tan, P.C., Yeo, S.H., Tan, Q.P., 2009. Surface roughness model for micro electrical discharge machining. *Proc. Inst. Mech. Eng. Part B J. Eng. Manuf.* 223, 279–287.

Lhiaubet, C., Meyer, R.M., 1981. Method of indirect determination of the anodic and cathodic voltage drops in short high-current electric discharges in a dielectric liquid. *J. Appl. Phys.* 52, 3929–3934.

Maradia, U., Hollenstein, C., Wegener, K., 2015. Temporal characteristics of the pulsed electric discharges in small gaps filled with hydrocarbon oil. *J. Phys. D. Appl. Phys.* 48, 55202.

Maradia, U., Wegener, K., 2015. EDM modelling and simulation. In: M.P. Jahan (ed) *Electrical Discharge Machining (EDM) Types, Technologies and Applications.* Nova Science Publishers, Inc., New York, pp. 67–121.

Mastud, S.A., Kothari, N.S., Singh, R.K., Joshi, S.S., 2015. Modeling debris motion in vibration assisted reverse micro electrical discharge machining process (R-MEDM). *J. Microelectromech. Syst.* 24, 661–676.

Mujumdar, S.S., Curreli, D., Kapoor, S.G., Ruzic, D., 2013. A model of micro electro-discharge machining plasma discharge in deionized water. *J. Manuf. Sci. Eng.* 136, 031011.

Mujumdar, S.S., Curreli, D., Kapoor, S.G., Ruzic, D., 2014. Modeling of melt-pool formation and material removal in micro-electrodischarge machining. *J. Manuf. Sci. Eng.* 137, 031007.

Murali, M.S., Yeo, S.-H., 2005. Process simulation and residual stress estimation of micro-electrodischarge machining using finite element method. *Jpn. J. Appl. Phys.* 44, 5254.

Nagahanumaiah, J.R., Glumac, N., Kapoor, S.G., Devor, R.E., 2009. Characterization of plasma in micro-EDM discharge using optical spectroscopy. *J. Manuf. Process.* 11, 82–87.

Pandey, P.C., Jilani, S.T., 1986. Plasma channel growth and the resolidified layer in EDM. *Precis. Eng.* 8, 104–110.

Patel, M.R., Barrufet, M.A., Eubank, P.T., DiBitonto, D.D., 1989. Theoretical models of the electrical discharge machining process. II. The anode erosion model. *J. Appl. Phys.* 66, 4104–4111.

Revaz, B., Witz, G., Flükiger, R., 2005. Properties of the plasma channel in liquid discharges inferred from cathode local temperature measurements. *J. Appl. Phys.* 98, 113305.

Roy, T., Datta, D., Balasubramaniam, R., 2018. Numerical modelling, simulation and fabrication of 3-D hemi-spherical convex micro features using reverse micro EDM. *J. Manuf. Process.* 32, 344–356.

Schoenbach, K., Kolb, J., Xiao, S., Katsuki, S., Minamitani, Y., Joshi, R., 2008. Electrical breakdown of water in microgaps. *Plasma Sources Sci. Technol.* 17, 024010.

Shabgard, M., Ahmadi, R., Seyedzavvar, M., Oliaei, S.N.B., 2013. Mathematical and numerical modeling of the effect of input-parameters on the flushing efficiency of plasma channel in EDM process. *Int. J. Mach. Tools Manuf.* 65, 79–87.

Shao, B., Rajurkar, K.P., 2013. Micro-EDM pulse energy distribution ratio determination. *The 8th International Conference on MicroManufacturing (ICOMM 2013)*, Singapore.

Shao, B., Rajurkar, K.P., 2015a. Modelling of the crater formation in micro-EDM. *Procedia CIRP.* 33, 376–381. doi: 10.1016/j.procir.2015.06.085.

Shao, B., Rajurkar, K.P., 2015b. Modelling and simulation of the crater formation process in micro-EDM. Procedia CIRP. 33, 376–381.

Singh, H., 2012. Experimental study of distribution of energy during EDM process for utilization in thermal models. *Int. J. Heat Mass Transf.* 55, 5053–5064.

Snoyes, R., Van Dijck, F., 1971. Investigations of EDM operations by means of thermo mathematical models. *Ann. CIRP* 20, 35.

Somashekhar, K.P., Panda, S., Mathew, J., Ramachandran, N., 2013. Numerical simulation of micro-EDM model with multi-spark. *Int. J. Adv. Manuf. Technol.* 76, 83–90. doi: 10.1007/s00170-013-5319-9.

Subbu, S.K., Dhamodaran, S., Ramkumar, J., 2012. Microelectric discharge plasma: Characterization and applications. *Mater. Manuf. Process.* 27, 1208–1212.

Tan, P.C., Yeo, S.H., 2008. Modelling of overlapping craters in micro-electrical discharge machining. *J. Phys. D. Appl. Phys.* 41, 205302.

Tao, J., Ni, J., Shih, A.J., 2012. Modeling of the anode crater formation in electrical discharge machining. *J. Manuf. Sci. Eng.* 134, 11002.

Van Dijck, F., 1973. Physico-mathematical analysis of the EDM process. Doctoral dissertation, PhD Thesis, Katholieke University, Leuven, Netherlands.

Van Dijk, F., Snoeys, R., 1971. Thermo-mathematical analysis of electro-discharge machining operations. *Proceeding of CNTN Conference*, Timisoara, Roumania.

Wang, Y.K., Xie, B.C., Wang, Z.L., Peng, Z.L., 2011. Micro EDM deposition in air by single discharge thermo simulation. *Trans. Nonferrous Met. Soc. China*, English Ed. 21, S450–S455.

Watson, P.K., 1985. Electrostatic and hydrodynamic effects in the electrical breakdown of liquid dielectrics. *IEEE Trans. Electr. Insul.* EI-20, 395–399.

Xia, H., Kunieda, M., Nishiwaki, N., 1996. Removal amount difference between anode and cathode in EDM process. *IJEM* 1, 45–52.

Yang, X., Guo, J., Chen, X., Kunieda, M., 2011. Molecular dynamics simulation of the material removal mechanism in micro-EDM. *Precis. Eng.* 35, 51–57.

Yeo, S.H., Kurnia, W., Tan, P.C., 2007. Electro-thermal modelling of anode and cathode in micro-EDM. *J. Phys. D. Appl. Phys.* 40, 2513–2521.

Zahiruddin, M., Kunieda, M., 2010. Energy distribution ratio into micro EDM electrodes. *J. Adv. Mech. Des. Syst. Manuf.* 4, 1095–1106.

Zahiruddin, M., Kunieda, M., 2012. Comparison of energy and removal efficiencies between micro and macro EDM. *CIRP Ann. Manuf. Technol.* 61, 187–190.

Zhang, F., Gu, L., Zhao, W., 2015. Study of the gaussian distribution of heat flux for micro-EDM. *ASME 2015 International Manufacturing Science and Engineering Conference*, American Society of Mechanical Engineers, North Carolina, V001T02A024

Zhang, Y., Liu, Y., Shen, Y., Li, Z., Ji, R., Wang, F., 2013. A new method of investigation the characteristic of the heat flux of EDM plasma. *Procedia CIRP* 6, 450–455.

10

Tool Wear Modeling and Compensation Methods

10.1 Introduction

Fabricating microfeatures using micro electro discharge machining (EDM) is possible because of the melting and vaporization of workpiece material by electrical discharges. However, the same discharges are also responsible for tool erosion, which is commonly known as tool wear. A part of the thermal energy is also transferred to the tool electrode, and craters are formed on the electrode surface due to material ablation. The erosion will eventually reduce the tool electrode length and distorts the tool electrode geometry. These changes will be reflected as deviations from the desired shape and dimensions in the fabricated microfeature. Tool electrode wear is considered as a primary challenge to implement micro EDM process as a precision micromachining technique. The tool wear in micro EDM can be divided into lengthwise wear and corner wear. The wear that reduces the length of the microelectrode is called lengthwise wear, and the wear that causes rounding at the corners is called corner wear as shown in Figure 10.1.

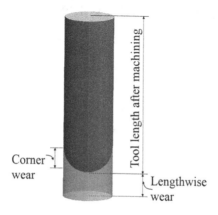

FIGURE 10.1
Wear characteristics of a micro tool in EDM.

10.2 The Principle of Tool Wear Formation in Micro EDM

Even though the mechanism of tool wear formation is the same in all micro EDM variants, the wear characteristics may change according to the micro EDM variant.

10.2.1 Micro EDM Drilling

In micro EDM drilling, a simple-shaped tool is rotated and plunged towards the workpiece to drill a hole. During this process, the bottom surface of the tool is constantly engaged with the workpiece electrode rather than the cylindrical surface as shown in Figure 10.2. In the primary machining region (tool electrode bottom), the electric field intensity will be maximum at the rim of the electrode at the starting of the machining process, as shown in Figure 10.3a. Due to this reason, most of the sparks will be concentrated in this area. Eventually, the tool material is melted and evaporated from the rim area and the tool shape changes. The rounding of the rim surface distorts the electric field, and it is concentrated towards the center as shown in Figure 10.3a. As the probability of discharge formation at a point depends on the intensity of the electric field, the area near the center will be responsible for further material removal (Li et al., 2014). Figure 10.3b,c shows the change in the tool electrode shape during the drilling process (Ekmekci et al., 2009). Interestingly, the tool shape remains unchanged after the initial corner rounding, and the bottom surface of the hole becomes curved in shape.

10.2.2 Micro EDM Milling

In micro EDM milling, the cylindrical surface is largely responsible for machining, and this causes a higher side wear compared to micro EDM drilling as shown in Figure 10.4a,b. Compared to EDM drilling, the tool

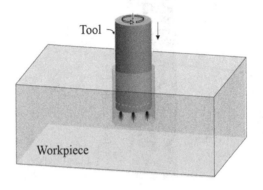

FIGURE 10.2
Tool wear in micro EDM drilling.

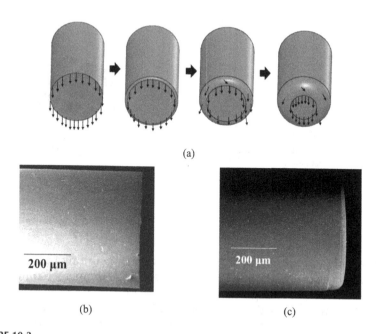

(a)

(b)

(c)

FIGURE 10.3
(a) Change in electrical field intensity during micro EDM drilling, and side view of a micro EDM tool (b) before drilling and (c) after drilling.

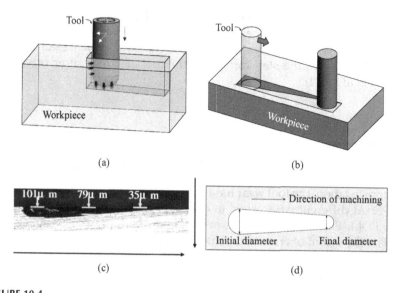

(a)

(b)

(c)

(d)

FIGURE 10.4
(a, b) Tool wear formation, (c) depth change in microchannel during micro EDM milling, and (d) change in the width of the microchannel due to continuous tool erosion (Yan et al., 2009).

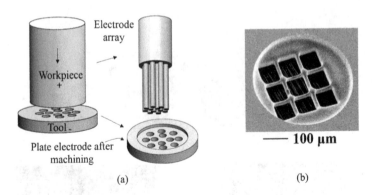

FIGURE 10.5
(a) Tool wear formation in reverse micro EDM and (b) eroded tool electrode (Yi et al., 2008).

electrode shape is not retained after initial corner rounding but undergoes a continuous change. Figure 10.4c shows the change in microchannel depth during micro EDM milling due to tool erosion (Yan et al., 2009). As the tool moves along the tool path, the shape changes that affect the dimensional accuracy of the machined microfeature are shown in Figure 10.4d.

10.2.3 Reverse EDM

During reverse EDM, the tool electrode plate experiences substantial erosion from the tool surface and eventually creates a step on the entrance face of the plate electrode. The lost material from the plate electrode surface will be responsible for the reduced height of the electrode array. Figure 10.5a shows the reverse EDM process with tool wear progression on plate electrode, and Figure 10.5b shows the step formed in the tool electrode after machining (Mastud et al., 2012).

10.3 Parameters to Quantify the Tool Electrode Wear

For compensation, the tool wear has to be quantified first. Different researchers defined different parameters to quantify the effect of tool wear during micro EDM. Figure 10.6 gives an overview of the tool wear parameters to calculate the extent of error in the machined profile.

10.3.1 Linear Wear

The difference in tool electrode length before and after machining is defined as linear wear. This will be replicated as the difference in intended depth

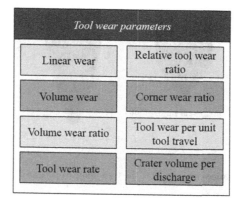

FIGURE 10.6
Wear parameters used by different researchers to quantify tool wear in EDM.

and actual depth during micro EDM drilling. However, this parameter fails to quantify the effect of corner wear.

10.3.2 Volume Wear Ratio

The volumetric wear is the ratio of the eroded volume of the electrode to the eroded volume of the workpiece. The volumetric wear or wear ratio mainly depends on parameters like current, voltage, and pulse-on time, electrode materials, and dielectric fluid properties. The simplified equation for finding the volumetric wear ratio, formulated by Pham et al. (2007), is given by Eq. (10.1):

$$v = \frac{V_e}{V_h} = \frac{D_e^2}{D_h^2}\left(\frac{1}{\dfrac{Z}{t_e}-1}\right) \tag{10.1}$$

where v is the volumetric wear ratio, V_e is the eroded volume from the tool electrode, V_h is the volume of the hole, t_e is the length erode from the tool, D_e is the diameter of the tool, D_h is the diameter of the hole, and Z is the targeted depth.

10.3.3 Tool Wear Rate

$$\text{Tool wear rate} = \frac{\text{Initial weight of the tool} - \text{final weight of the tool}}{\text{Machining time}} \tag{10.2}$$

Or

$$\text{Tool wear rate} = \frac{\text{Initial volume of the tool} - \text{final volume of the tool}}{\text{Machining time}} \tag{10.3}$$

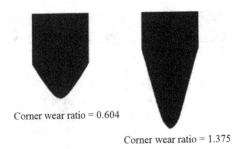

Corner wear ratio = 0.604

Corner wear ratio = 1.375

FIGURE 10.7
Change in corner wear ratio with respect to the wear length and tool diameter (Yan et al., 2009).

10.3.4 Relative Tool Wear Ratio

$$\text{Relative tool wear ratio} = \frac{\text{Material removal rate of the workpiece}}{\text{Tool wear rate}} \quad (10.4)$$

10.3.5 Corner Wear Ratio

The ratio between the length of the corner wear and the electrode diameter is known as corner wear ratio (Yan et al., 2009). Figure 10.7 shows the change in corner ratio as the tool transforms into a conical shape.

10.3.6 Tool Wear per Unit Tool Travel

It is the ratio between the tool wear and the total tool travel distance that is used to quantify the erosion of the tool electrode during blind-hole drilling (Malayath et al., 2019).

10.3.7 Tool Wear per Discharge

The volume of material eroded from the tool per discharge can be calculated by dividing the volumetric wear of the tool with the number of normal discharges (Nguyen et al., 2015).

10.4 Assessment of Tool Wear

After machining, the change in tool length and diameter may be measured using an optical microscope or a scanning electron microscope. However, this is a time-consuming process. Moreover, the wear compensation procedure cannot be applied once the workpiece or tool is removed from the fixtures. So, the tool wear has to be assessed by on-machine measurements

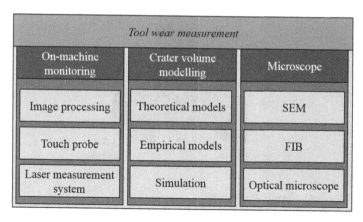

FIGURE 10.8
Different methods used for tool wear assessment during micro EDM.

or theoretical models. Figure 10.8 shows the different methods used for the assessment of tool wear.

10.4.1 On-Machine Measurement of Tool Wear

To quantify the tool wear using offline methods, an on-machine measurement system has to be realized. This will help to measure the change in length and shape of the tool electrode in frequent intervals to indirectly estimate the extent of errors in the machined profile.

10.4.1.1 Touch Probe Method

A touch probe attached to the machine tool is used to measure the change in electrode length using electrode sensing methods (Kaneko and Tsuchiya, 1988). The tool electrode is allowed to touch the workpiece surface or any other reference point in the working zone. The coordinates where short circuit occurs between the tool electrode and the reference point are recorded. The process is repeated after frequent intervals to find out the change in electrode length during each step. However, a small amount of erosion during contact may cause measurement errors (Bissacco et al., 2010a).

10.4.1.2 Laser Measurement System

A laser measurement system can be utilized to understand the change in length and diameter of the tool electrode wear. A laser range sensor consists of a laser projected array, and a linear detector array is used for on-machine measurement of the tool electrode. The high-resolution array is capable of measuring the diameter and length of the rod, which can be further used to calculate the amount of tool electrode wear (Mizugaki, 1996; Bissacco et al.,

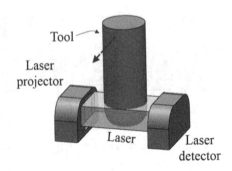

FIGURE 10.9
Principle of the laser measurement system.

2010a). Bissacco et al. (2010a) compared laser measurement system and touch probe method in terms of measured volumetric wear and linear wear. The standard uncertainty in touch probe method is found to be slightly higher than the laser measurement method. However, the accuracy of the laser measurement system largely depends on the resolution and repeatability of the machine tool positioning systems. Venugopal et al. (2014) used this method for drilling blind holes using micro EDM. Figure 10.9 shows the basic arrangement of a laser measurement system (Mizugaki, 1996).

10.4.2 Image Processing Method

Using a camera and an image-processing software, the image of the tool before and after machining can be captured (Malayath et al., 2019). The tool image is then processed with an algorithm to calculate the length of the electrode based on a pixel per unit length value of the captured image. Figure 10.10a shows the setup for on-machine measurement of tool wear using image-processing method. Figure 10.10b shows the processed tool image. The coordinates of pixels at different points in the tool image are used to find out the difference in tool length. Guo et al. (2006) conducted tool wear study using image processing, and Yan et al. (2009) introduced the first tool wear compensation method for micro EDM milling based on this strategy. Yan and Lin (2011) used image processing-based tool wear compensation for multicut EDM milling.

10.4.3 Modeling

Tool wear crater volume can be calculated by electrothermal models of micro EDM process (Yeo et al., 2007). The boundary of the crater is estimated to get the diameter and depth of the hole. The crater volume is numerically calculated and is used to find out the tool wear ratio. Aligiri et al. (2010) also used the electrothermal model for tool wear monitoring.

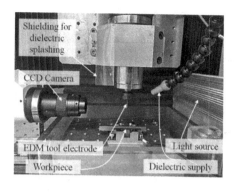

(a)   (b)

FIGURE 10.10
(a) Setup and (b) processed tool electrode image in image processing-based tool wear compensation system (Malayath et al., 2019).

10.5 Tool Wear Compensation

The calculated tool wear is compensated with various strategies. According to the method of compensation and tool wear monitoring, the compensation strategies are broadly classified as offline and online methods. In offline methods, periodic measurements of the tool electrode length and diameter are carried out to predict the wear. In online compensation, the tool wear is assessed real time during the machining process by analyzing the voltage and current signals. Figure 10.11 shows the general classification of tool wear compensation strategies.

10.6 Offline Tool Wear Compensation

In offline compensation methods, the tool path trajectory is modified before machining using certain logic to minimize the effects of tool electrode wear. Characteristics of the tool wear parameters at specific machining conditions are analyzed separately using experimental analysis or by on-machine monitoring of tool condition. The wear parameters (relative tool wear ratio, tool wear length, tool wear per unit tool travel, etc.) are used to control the tool electrode motion. As the wear measurement and the tool path generation are done before starting machining, this type of compensation method is considered to be in the offline mode. Tool wear trend quantified is used to predict the tool wear for subsequent machining operations. According to the

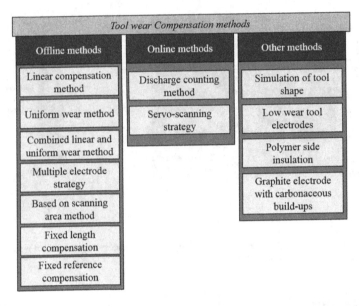

FIGURE 10.11
Classification of the tool wear compensation strategies.

logic used to modify the tool path, offline compensation methods are classified as follows:

1. Linear compensation method (LCM)
2. Uniform wear method (UWM)
3. Combined linear uniform method (CLU)
4. Multiple electrode strategies
5. Fixed length compensation method (FLC).

10.6.1 Linear Compensation Method

In this method, linear wear is compensated with an extra vertical feed provided continuously or at frequent intervals. The anticipated tool wear value calculated from relative tool wear ratio (found out by experiments or on-machine measurements) is used as a compensation value. This value is added to the final tool travel distance (for drilling) as shown in Figure 10.12a (Yeo et al., 2007). It can also be added at specific intervals by keeping the electrode feed depth to the tool travel distance constant (for milling) as shown in Figure 10.12b. This strategy of intermittent feed correction has the advantage of attaining intended depth involving less complicated wear calculation steps. Malayath et al. (2019) used LCM to drill a series of blind microholes on a tungsten carbide workpiece. Three-dimensional surfaces with straight walls can be machined by providing linear wear compensation. However,

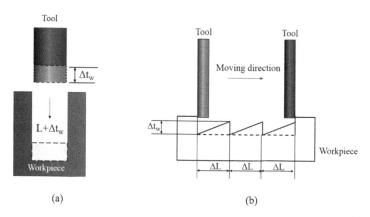

FIGURE 10.12
Linear compensation strategy for (a) micro EDM drilling and (b) micro EDM milling.

the capability of LCM to machine 3D complex surfaces is doubted by various researchers (Yu et al., 2010; Bleys et al., 2004). Moreover, the effect of corner wear during machining is not addressed while employing LCM. Lack of corner wear compensation reduces the shape accuracy of the machined microfeatures.

10.6.2 Uniform Wear Method

UWM corresponds to the method of converting the volume wear of the tool to linear wear by employing a layer-by-layer machining method. Yu et al. (1998b) found that by selecting appropriate layer thickness in layerwise EDM milling, the effect of corner wear can be nullified. When the tool path is designed based on UWM, the tool wear is visible only at the tool bottom surface, and the tool shape is recovered. The change in tool length can be compensated by adding the eroded length to the thickness of the next layer. The effectiveness of this method largely depends on the tool path used. The tool path should ensure uniform wear of the tool along the layers. Yu et al. (1998a) studied the effect of tool path, electrode wear ratio, and tool electrode size on the tool wear compensation. Dimov et al. (2003) introduced a computer aided design and manufacturing (CAD/CAM) tool for tool path generation based on adaptive slicing method. The increment applied on the ith layer (ΔZ_i) of the machining process is given by Eq. (7.5):

$$\Delta Z_i = L_{W_i}\left(1 + \frac{vS_{w_i}}{S_e}\right) \tag{10.5}$$

where
L_{W_i} = thickness of the ith layer
v = volumetric wear ratio

S_{w_i} = area of the ith layer
S_e = cross-sectional area of the electrode.

The tool path should be designed with the following characteristics (Yu et al., 2010):

- Layer-by-layer machining for nullifying the effect of corner wear
- To-and-fro scanning to minimize the inclination of the machined surface
- Overlapping of the tool path to reduce the surface roughness and avoiding nonmachined areas
- Alternative machining of the boundary and center part of the microfeature.

If the layer thickness selected is smaller than the spark gap, the chance of side sparking is substantially reduced. However, the inclination of the machined layer cannot be completely removed while using UWM. Moreover, if the compensation length is large, the thickness of the subsequent layer will be very high, which will affect the machining stability.

10.6.3 Combined Linear Uniform Wear Method

In the CLU method, instead of adding the tool wear compensation length at the starting of the layer, it is evenly divided and fed in equal intervals in a layer. This can be considered as a combination of UWM and LCM. The principle of CLU method is explained in Figure 10.13. Yu et al. (2010) reported the improvement of employing a hybrid method in reducing the tool wear and improving the surface quality.

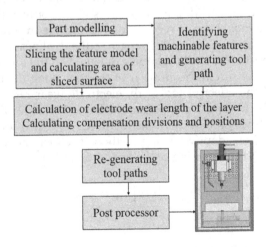

FIGURE 10.13
CAD/CAM module for tool wear compensation.

10.6.4 Multiple Electrode Strategies

Instead of using a single electrode to complete the tool path during micro EDM milling, multiple electrodes can be used with frequent dressing cycles. After a specific period, the tool electrode is pulled back and dressed with an electro discharge grinding (EDG) arrangement equipped within the machine tool. Flattening of the tool bottom surface reduces the effects of corner wear. After dressing operation, the coordinates of the tool electrode concerning the workpiece surface has to be modified to avoid positioning errors. The dressing operation and surface point resetting operation added to the total machining cycle increase the machining time. For making this process more effective, the number of passes and dressing cycles has to be optimized. Karthikeyan et al. (2011) studied the tool wear in a single pass micro EDM milling using CAD-based simulation method and used the results to compensate the tool wear in the multipass machining process. Regression analysis is used to generalize the model for all machining conditions. The feasibility of using a multistep electrode for multipass EDM milling analyzed by Karthikeyan et al. (2014) is shown in Figure 10.14. Shukla et al. (2013) fabricated long serpentine microchannels and T-channels using multiple pass approach.

10.6.5 Based on the Scanned Area Method

While using CLU, the accuracy mostly depends on the position of the compensation point. To enhance the efficiency of CLU method, the electrode wear length is divided by the resolution of the axis positioning system (Li et al., 2013). From this, the number of compensation points is calculated. Area of each sliced layer is then divided to a finite number of grids in the X-Y plane as shown in Figure 10.15. During machining, the grids are scanned continuously. The compensation value is calculated from the number of scanned grids and the grid area. This method shows better performance in terms of relative tool wear and material removal rate compared to the CLU method.

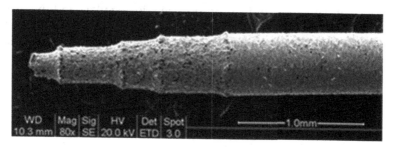

FIGURE 10.14
Multistep micro tool developed for micro EDM milling (Karthikeyan et al., 2014).

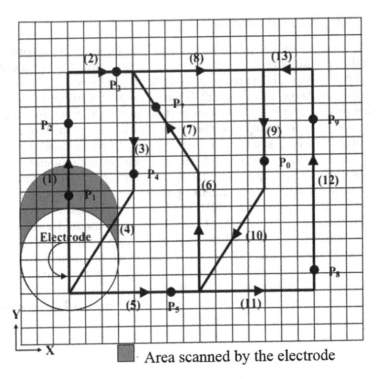

FIGURE 10.15
Principle of the grid based on the scanned area method (Li et al., 2013).

10.6.6 Fixed Length Compensation

To increase the layer thickness in UWM, an FLC strategy is introduced by Pei et al. (2013). The feasibility of increasing the layer thickness with the help of a more precise theoretical model of electrode wear, and the machined surface is analyzed. For this, the compensation point is fixed with the help of a model for tool thickness. However, the increase in layer thickness makes the tool conical in shape, which brings dimensional errors during 3D machining. The conical end demands postprocessing during plane milling to avoid fluctuations in the surface profile. Pei et al. (2016) introduced plane milling methods using FLC with the help of tubular electrodes. The wear characteristics of machining with FLC and tubular electrodes are studied by Pei et al. (2018). The FLC model is modified to get higher machining performance using large layer thickness in micro EDM milling (Zhang et al., 2012).

10.6.7 Fixed Reference Compensation

In this, the workpiece surface coordinate values are updated after machining each layer by establishing electrical contact with a fixed reference point

(Wang and Dong, 2007). The change in the origin points will be a measure of the tool wear. Simplicity is the main feature of this strategy. Many researchers have employed this method for fabricating various microfeatures (Modica et al., 2014; Hang et al., 2006; Modica et al., 2011). However, the accuracy of this method highly depends on the resolution and repeatability of the machine tool (Kar and Patowari, 2018).

10.6.8 The Drawback of Offline Compensation Methods

- Calculation of anticipated wear from periodical measurements requires stopping the machining process frequently. This increases the machining time considerably.
- Offline tool wear monitoring and compensation requires precise knowledge about the microfeature geometry to calculate the volume of material removed from the workpiece (Bleys et al., 2004).
- Errors in the calculation of the predicted wear add up as the layerwise machining progresses and the cumulative effect produces dimensional errors in the final microcomponent.

10.7 Online Tool Wear Compensation

Online compensation methods refer to the method of compensating the wear without interrupting the machining process. Even though the inputs for the calculations of tool wear parameters demand some assistance from offline tool monitoring methods or experimental data, the tool wear during machining is calculated real time.

10.7.1 Discharge Counting Method

To assess the tool wear rate or tool wear length per layer in real time, direct measuring methods like optical measuring system cannot be used, as the visibility is severely compromised in the presence of dielectric fluid. So, the wear-sensing method has to be indirect. As the voltage and pulse characteristics change during discharging, analyzing the pulses give an idea about the status of the tool electrode. Using an efficient pulse discrimination method, the discharges can be divided into normal discharges (useful discharges), arcs, and short circuits (Liao et al., 2008). The pulse discrimination system works on different principles, based on different voltage and current parameters, which is elaborately discussed in Chapter 2.

Tool wear per discharge value is calculated using pilot experiments, on-machine monitoring systems, previous experimental data or via theoretical modeling of the micro EDM process. Considering normal discharges as the reason behind material erosion in the workpiece and tool, the number of normal discharges is found using discharge counting methods. To calculate the amount of tool wear during machining, this normal discharge count is then multiplied with the tool wear per discharge value. The eroded length thus found is then compensated accordingly. However, the discharges in micro EDM do not always carry the same energy. Assigning the same crater volume per discharge value for all the sparks will result in error in tool wear calculations. Figure 10.16 shows the necessary steps in the online tool wear compensation method (Nirala and Saha, 2017a). Online wear compensation methods based on discharge counting are differentiated in terms of the methods used for assessing the tool wear per unit discharge and pulse discrimination systems. Real-time tool wear calculations based on the assumption of isoenergetic discharges are done by Jung et al. (2007, 2008), Bissacco et al. (2013), etc. More than 11% error in

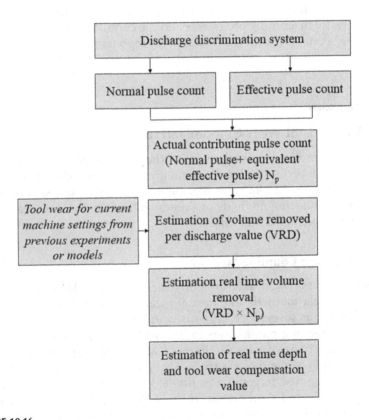

FIGURE 10.16
General steps in online tool wear compensation based on discharge counting servo scanning method.

the tool wear calculations while applying isoenergetic assumption has been reported (Jung et al., 2007) when the discharge energy is high. Bissacco et al. (2011) used a statistical distribution of discharges and mapped it with population tool wear per discharge to calculate the tool wear during micro EDM milling. The results are compared with tool wear measured using offline tool monitoring systems, and a maximum error of 2 μm is reported. Roundness factor is incorporated in the wear compensation value calculations for micro EDM drilling by Nirala and Saha (2017a), and less than 4% error in the depth of the hole is reported.

10.7.2 Servo Scanning Method

A method to compensate the tool wear by efficient control over the discharge gap during layer-by-layer machining of microfeatures is proposed by Li et al. (2007). Analyzing the voltage and current signal, the servo feed is controlled along the tool path to get a constant discharge gap. The steps include modeling of the 3D microstructure, generating numerical control (NC) codes and fabricating the microstructure by point-to-point and layer-by-layer servo scanning (Tong et al., 2014). While employing this method, the feed of the microelectrode is adapted to the changing electrode wear. Moreover, the wear parameter inputs from experimental data are not needed for tool wear compensation as the process depends on the principle of automatic wear compensation using servo control of the discharge gap. However, the cumulative effects of the depth errors (from workpiece positioning, scanning paths, etc.) introduce dimensional inaccuracies and uneven surface during the contour scanning process. A layer depth constraint algorithm and S-curve accelerating algorithm are used to reduce overcutting and errors due to insufficient machining (at the starting and ending of each layer) (Tong et al., 2014). Using servo scanning EDM for roughing and finishing operations, the depth errors were limited to 2 μm, and the machining efficiency is improved by 2.4 times (Tong et al., 2016).

10.7.3 Drawbacks of Online Compensation Methods

- Errors in the determination of tool wear per discharge results in error in layer thickness calculations. These errors may be accumulated and may lead to depth overcut during milling and drilling processes.
- The efficiency of discharge counting method largely depends on the pulse discrimination algorithm.
- The isoenergetic assumption introduces errors in the wear calculation as the discharge energy depends on pulse duration and type of pulse generator used.

10.8 Other Methods for Prediction, Correction, and Reduction of Tool Wear

- Puthumana et al. (2017) studied the effect of error in calculating wear per discharge on the machined profile using offline simulation. The effect of error propagation during machining is studied, and methods to correct the tool wear per discharge values are suggested.

- The effect of discharge gap, corner wear, and tool wear on the profile of the machined surface is analyzed with the help of a virtual tool electrode simulation (Nguyen et al., 2013). Profile error due to the discharge gap and tool wear can be reduced by selecting appropriate offset distance during machining. The virtual electrode simulation will help to calculate optimum offset distance.

- Li et al. (2012) conducted a simulation study of micro EDM process by analyzing the change in electric field intensity concerning the progression of tool wear.

- A virtual EDM simulator to predict the tool shape during micro EDM drilling operation is realized by Heo et al. (2009). A Z-map algorithm is used to represent the tool wear progression during machining. The simulation tool can be applied for tool path generation and tool wear prediction during micro EDM.

- A two-dimensional geometric prediction model of micro EDM drilling process is developed by Jeong and Min (2007). Normal spark count, crater volume per discharge, and tool wear ratio are used to model the machining process. This simulation tool is used to assist the offline compensation methods. The tool wear compensation error is reduced to 2.9%.

- Electrically conductive chemical vapor deposition (CVD) diamond is used as a tool material for micro EDM to reduce the tool wear (Suzuki et al., 2006). High thermal stability of the diamond and accelerated deposition of carbon on the surface is responsible for low tool wear.

- Cu-ZrB$_2$ composite-coated electrodes fabricated by electrodeposition process are used for micro EDM as low-wear tool electrodes (Yuangang et al., 2009). The coating serves as a protective sheath against the erosion of the side surface of the tool.

- Feasibility of using polymer side insulation of the micro EDM electrodes to reduce side wear is studied by Hu et al. (2011) and Ferraris et al. (2013). The achievable aspect ratio is improved by 30%, and the tool wear is lowered by half using insulated tool electrodes.

- Study on the formation of carbonaceous buildups on the graphite electrode and development of a method to achieve low wear during machining by applying a combination of long and short pulses is carried out by Maradia et al. (2015a, b).

Table 10.1 consolidates the major research works in the EDM tool wear compensation and classifies the findings based on the types of methods used for tool wear monitoring and correction.

TABLE 10.1

Summary of Research Works in Tool Wear Compensation in Micro EDM Process

Authors	Online/ Offline	Tool Wear Parameter	Method of Assessing Tool Wear	Micro EDM Variant and Comments on the Tool Compensation Method
Jeswani (1979)	Offline	Eroded volume from tool electrode	Dimensional analysis of tool wear	*Die sinking* An analytical method to calculate tool wear
Yu et al. (1998b)	Offline	Relative electrode wear ratio	–	*Milling* Introduction of UWM method for wear compensation
Yu et al. (1998a)	Offline	Volume electrode wear ratio	–	*Milling* Experimental analysis by varying volume electrode wear ratio
Bleys et al. (2002)	Offline simulation and online sensing	Volumetric wear ratio, constant volumetric wear per discharge	Previous experimental results	*Milling*
Bleys et al. (2004)	Offline Online Combined	Linear wear, eroded volume	Touch probe, discharge counting	*Milling* Layer by layer (UWM)
Dimov et al. (2003)	Offline	Volumetric wear ratio, area of layer	Previous experimental results	*Milling* UWM with adaptive slicing method
Narasimhan et al. (2005)	Offline	Electrode wear ratio	Previous experimental results	*Milling* Profile simulation
Guo et al. (2006)	Offline	Electrode diameter	Image processing	*Milling* Wear measurement only
Pham et al. (2007)	Offline	Volumetric wear ratio, corner wear ratio	Tool wear calculation from scanning electron microscopy (SEM) images	*Drilling*

(Continued)

TABLE 10.1 (*Continued*)

Summary of Research Works in Tool Wear Compensation in Micro EDM Process

Authors	Online/ Offline	Tool Wear Parameter	Method of Assessing Tool Wear	Micro EDM Variant and Comments on the Tool Compensation Method
Yeo et al. (2007)	Offline	Crater volume, tool wear ratio	Theoretical model	*Drilling* Linear wear compensation
Jeong and Min (2007)	Offline	Workpiece crater area, areal tool wear ratio, actual tool crater area	Workpiece crater volume by single spark experiments	*Drilling* Simulation of tool and workpiece shape after machining by finding crater volume and sparking points
Jung et al. (2008)	Online	Moving average pulse frequency	Pulse frequency sensing	*Milling* Tool wear compensation by real-time gap control based on the desired pulse frequency
Yan et al. (2009)	Offline	Frontal and corner wear	Image processing	*Milling* Layer-by-layer machining with linear tool wear compensation
Heo et al. (2009)	Offline	Workpiece crater volume, volumetric tool wear ratio	Previous experimental results	*Milling* Simulation of workpiece and tool shape after machining using Z-map algorithm
Yu et al. (2010)	Offline	Relative volumetric electrode wear ratio	Previous experimental results	*Milling* Combination of linear compensation and uniform wear compensation (CLU)
Aligiri et al. (2010)	Online	Workpiece crater volume, number of normal discharges	Electrothermal model Discharge counting	*Drilling* Linear wear compensation
Bissacco et al. (2010b)	Online/ offline (for reference data generation)	The volume of electrode wear per discharge, number of normal discharges	Pulse discrimination Mapping number of pulses with the channel volume	*Microgrooving* To study the tool wear per discharge for different discharge energy index

(Continued)

TABLE 10.1 (*Continued*)

Summary of Research Works in Tool Wear Compensation in Micro EDM Process

Authors	Online/ Offline	Tool Wear Parameter	Method of Assessing Tool Wear	Micro EDM Variant and Comments on the Tool Compensation Method
Bissacco et al. (2011)	Online Offline (for verification)	Population tool wear per discharge, number of normal discharges	Discharge counting, laser measurement system	*Micromilling* Calculation of error in tool wear per discharge
Yan and Lin (2011)	Offline	Linear wear, corner wear ratio	Image processing	*Milling* Multicut process with varying feed rate per layer
Bissacco et al. (2010a)	Offline	Linear wear, volumetric wear	Comparison of on-machine measurement methods	*Drilling*
Chen and Yang (2011)	Offline	Linear wear	Image processing, contact probe	*Milling*
Bissacco et al. (2013)	Online	Material removed per discharge	Touch probe Discharge counting	*Milling* Layer-by-layer machining (UWM)
Nguyen et al. (2013)	Offline	Volumetric wear ratio, linear tool wear per layer	Varying the wear parameters manually for simulation	*Milling* Calculation of profile error during layer-by-layer machining
Li et al. (2013)	Offline	Electrode wear ratio	Previous experimental results	*Milling* Based on scanned area, modification of CLU by slicing the tool path into a large number of grids
Venugopal et al. (2014)	Offline	Linear wear	Laser measurement system	*Drilling* Blind hole
Tong et al. (2014)	Offline Online	Laminated thickness, wear coefficient	Preliminary experiments for getting wear parameters, real-time servo control of spark gap	*Milling* Layer-by-layer machining with layer depth constrained algorithm and S-curve accelerating algorithm

(Continued)

TABLE 10.1 (*Continued*)

Summary of Research Works in Tool Wear Compensation in Micro EDM Process

Authors	Online/ Offline	Tool Wear Parameter	Method of Assessing Tool Wear	Micro EDM Variant and Comments on the Tool Compensation Method
Zhang et al. (2015)	Offline	Workpiece crater diameter	Previous experimental results	*Milling* Simulation of tool and workpiece shape during milling with FLC using a cylindrical tool
Nguyen et al. (2015)	Online Offline (for generating tool wear per discharge data)	Tool wear per discharge, number of normal discharge pulses	Discharge counting (lab view), touch probe	*Milling* UWM strategy for wear compensation
Puthumana et al. (2017)	Offline	Tool wear per discharge	Previous experimental results	*Milling* Profile simulation and correction Layer-by-layer machining
Nirala and Saha (2017a, b)	Online	Volume removed per discharge, number of effective discharges	Image processing, discharge counting	*Drilling* Target height-based compensation
Wang et al. (2017)	Online Offline (initial data generation)	Tool wear per discharge, number of normal pulses, tool feeding slope	On-machine wear, measurement (offline), discharge counting. Empirical model	*Milling* Linear tool wear compensation
Pei et al. (2009, 2013, 2016, 2018)	Offline	Volumetric wear ratio	Previous experimental results	*Milling* FLC method
Malayath et al. (2019)	Offline	Tool wear per unit tool travel	Image processing	*Drilling* Series of holes with constant depth using average forecasting algorithm

10.9 Conclusion

Tool wear is an inevitable phenomenon in micro EDM. To tackle this problem, a tool compensation method has to be established. Tool wear compensation starts with quantifying the wear using different parameters. Different methods are used to measure tool wear, including image processing, electrical sensing, laser measurement system, optical microscope, etc. Tool wear compensation methods are generally classified as online and offline compensation methods. Online methods calculate the tool wear real time and apply for the wear compensation without halting the machining process. However, offline methods use periodic measurements to forecast the tool wear. Apart from applying wear compensations, insulating the tool surface with different coatings and selecting low-wear materials for tool electrode are used to reduce tool wear in micro EDM.

References

Aligiri, E., Yeo, S.H., Tan, P.C., 2010. A new tool wear compensation method based on real-time estimation of material removal volume in micro-EDM. *J. Mater. Process. Technol.* 210, 2292–2303.

Bissacco, G., Hansen, H.N., Tristo, G., Valentincic, J., 2011. Feasibility of wear compensation in micro EDM milling based on discharge counting and discharge population characterization. *CIRP Ann. Manuf. Technol.* 60, 231–234.

Bissacco, G., Tristo, G., Hansen, H.N., Valentincic, J., 2013. Reliability of electrode wear compensation based on material removal per discharge in micro EDM milling. *CIRP Ann. Manuf. Technol.* 62, 179–182.

Bissacco, G., Tristo, G., Valentincic, J., 2010a. Assessment of electrode wear measurement in micro EDM milling. *Proceedings of the 7th International Conference on Multi-Material Micro Manufacture*, Bourg en Bresse and Oyonnax, France.

Bissacco, G., Valentincic, J., Hansen, H.N., Wiwe, B.D., 2010b. Towards the effective tool wear control in micro-EDM milling. *Int. J. Adv. Manuf. Technol.* 47, 3–9.

Bleys, P., Kruth, J.P., Lauwers, B., 2004. Sensing and compensation of tool wear in milling EDM. *J. Mater. Process. Technol.* 149, 139–146.

Bleys, P., Kruth, J.P., Lauwers, B., Zryd, A., Delpretti, R., Tricarico, C., 2002. Real-time tool wear compensation in milling EDM. *CIRP Ann. Manuf. Technol.* 51, 157–160.

Chen, S.-T., Yang, H.-Y., 2011. Study of micro-electro discharge machining (micro-EDM) with on-machine measurement-assisted techniques. *Meas. Sci. Technol.* 22, 65702.

Dimov, S., Pham, D.T., Ivanov, A., Popov, K., 2003. Tool-path generation system for micro-electro discharge machining milling. *Proc. Inst. Mech. Eng. Part B J. Eng. Manuf.* 217, 1633–1637.

Ekmekci, B., Sayar, A., Öpöz, T.T., Erden, A., 2009. Geometry and surface damage in micro electrical discharge machining of micro-holes. *J. Micromech. Microeng.* 19, 105030.

Ferraris, E., Castiglioni, V., Ceyssens, F., Annoni, M., Lauwers, B., Reynaerts, D., 2013. EDM drilling of ultra-high aspect ratio micro holes with insulated tools. *CIRP Ann. Manuf. Technol.* 62, 191–194.

Guo, R., Zhao, W., Li, G., Li, Z., Zhang, Y., 2006. A machine vision system for micro-EDM based on linux. *Third International Symposium on Precision Mechanical Measurements*, Urumqi, China. International Society for Optics and Photonics, 62803K.

Hang, G., Cao, G., Wang, Z., Tang, J., Wang, Z., Zhao, W., 2006. Micro-EDM milling of micro platinum hemisphere. *Proceeding of the 1st IEEE International Conference on Nano/Micro Engineered and Molecular Systems (IEEE-NEMS)*, Zhuhai, China, 579–584.

Heo, S., Jeong, Y.H., Min, B.K., Lee, S.J., 2009. Virtual EDM simulator: Three-dimensional geometric simulation of micro-EDM milling processes. *Int. J. Mach. Tools Manuf.* 49, 1029–1034.

Hu, M., Li, Y., Zhu, X., Tong, H., 2011. Influence of side-insulation film on hybrid process of micro EDM and ECM for 3D micro structures. *Adv. Mater. Res.* 232, 517–521.

Jeong, Y.H., Min, B.K., 2007. Geometry prediction of EDM-drilled holes and tool electrode shapes of micro-EDM process using simulation. *Int. J. Mach. Tools Manuf.* 47, 1817–1826.

Jeswani, M.L., 1979. Dimensional analysis of tool wear in electrical discharge machining. *Wear* 55, 153–161.

Jung, J.W., Jeong, Y.H., Min, B.-K., Lee, S.J., 2008. Model-based pulse frequency control for micro-EDM milling using real-time discharge pulse monitoring. *J. Manuf. Sci. Eng.* 130, 031106.

Jung, J.W., Ko, S.H., Jeong, Y.H., Min, B.K., Lee, S.J., 2007. Estimation of material removal volume of a micro-EDM drilled hole using discharge pulse monitoring. *Int. J. Precis. Eng. Man.* 8(4), 45–49.

Kaneko, T., Tsuchiya, M., 1988. Three-dimensional numerically controlled contouring by electric discharge machining with compensation for the deformation of cylindrical tool electrodes. *Precis. Eng.* 10, 157–163.

Kar, S., Patowari, P.K., 2018. Electrode wear phenomenon and its compensation in micro electrical discharge milling: A review. *Mater. Manuf. Process.* 33(14), 1491–1517.

Karthikeyan, G., Ramkumar, J., Dhamodaran, S., 2014. Block edg: Issues and applicability in multiple pass µED-milling. *Mach. Sci. Technol.* 18, 120–136.

Karthikeyan, G., Sambhav, K., Ramkumar, J., Dhamodaran, S., 2011. Simulation and experimental realization of µ-channels using a µED-milling process. *Proc. Inst. Mech. Eng. Part B J. Eng. Manuf.* 225, 2206–2219. doi: 10.1177/0954405411403359.

Li, J.Z., Xiao, L., Wang, H., Yu, H.L., Yu, Z.Y., 2013. Tool wear compensation in 3D micro EDM based on the scanned area. *Precis. Eng.* 37, 753–757. doi: 10.1016/j.precisioneng.2013.02.008.

Li, X.P., Wang, Y.G., Zhao, F.L., Wu, M.H., Liu, Y., 2014. Influence of high frequency pulse on electrode wear in micro-EDM. *Def. Technol.* 10, 316–320.

Li, X.P., Wu, M.H., Wang, H., Wang, Y.G., Liu, Y., 2012. A simulation study on rod electrode wear in micro-EDM. *Adv. Mater. Res.* 472–475, 2426–2429.

Li, Y., Tong, H., Cui, J., Wang, Y., 2007. Servo scanning EDM for 3D micro structures. *2007 First International Conference on Integration and Commercialization of Micro and Nanosystems*, Sanya, China. American Society of Mechanical Engineers, 1369–1374.

Liao, Y.S., Chang, T.Y., Chuang, T.J., 2008. An on-line monitoring system for a micro electrical discharge machining (micro-EDM) process. *J. Micromech. Microeng.* 18, 1–8.

Malayath, G., Katta, S., Sidpara, A.M., Deb, S., 2019. Length-wise tool wear compensation for micro electric discharge drilling of blind holes. *Measurement.* 134, 888–896.

Maradia, U., Boccadoro, M., Stirnimann, J., Kuster, F., Wegener, K., 2015a. Electrode wear protection mechanism in meso-micro-EDM. *J. Mater. Process. Technol.* 223, 22–33.

Maradia, U., Knaak, R., Dal Busco, W., Boccadoro, M., Wegener, K., 2015b. A strategy for low electrode wear in meso-micro-EDM. *Precis. Eng.* 42, 302–310.

Mastud, S., Singh, R.K., Joshi, S.S., 2012. Analysis of fabrication of arrayed micro-rods on tungsten carbide using reverse micro-EDM. *Int. J. Manuf. Technol. Manag.* 26, 176–195.

Mizugaki, Y., 1996. Contouring electrical discharge machining with on-machine measuring and dressing of a cylindrical graphite electrode. *Proceedings of the 1996 IEEE IECON: 22nd International Conference on Industrial Electronics, Control, and Instrumentation*. IEEE, Taipei, Taiwan, 1514–1517.

Modica, F., Guadagno, G., Marrocco, V., Fassi, I., 2014. Evaluation of micro-EDM milling performance using pulse discrimination. *ASME 2014 International Design Engineering Technical Conferences and Computers and Information in Engineering Conference*. American Society of Mechanical Engineers, V004T09A019–V004T09A019.

Modica, F., Marrocco, V., Copani, G., Fassi, I., 2011. Sustainable micro-manufacturing of micro-components via micro electrical discharge machining. *Sustainability* 3, 2456–2469.

Narasimhan, J., Yu, Z., Rajurkar, K.P., 2005. Tool wear compensation and path generation in micro and macro EDM. *J. Manuf. Process.* 7, 75–82.

Nguyen, M.D., Wong, Y.S., Rahman, M., 2013. Profile error compensation in high precision 3D micro-EDM milling. *Precis. Eng.* 37, 399–407.

Nguyen, V.Q., Duong, T.H., Kim, H.C., 2015. Precision micro EDM based on real-time monitoring and electrode wear compensation. *Int. J. Adv. Manuf. Technol.* 79, 1829–1838.

Nirala, C.K., Saha, P., 2017a. Precise μEDM-drilling using real-time indirect tool wear compensation. *J. Mater. Process. Technol.* 240, 176–189.

Nirala, C.K., Saha, P., 2017b. Toward development of a new online tool wear compensation strategy in micro-electro-discharge machining drilling. *Proc. Inst. Mech. Eng. Part B J. Eng. Manuf.* 231, 588–599.

Pei, J., Deng, R., Hu, D., 2009. Bottom surface profile of single slot and fix-length compensation method in micro-EDM process. *J. Shanghai Jiaotong Univ.* 43(1), 42–46.

Pei, J., Liu, Y., Zhu, Y., Zhang, L., Zhuang, X., Wu, S., 2018. Machining strategy and key problems for 3D structure of micro-EDM by fix-length compensation method with tubular electrodes. *Procedia CIRP* 68, 802–807.

Pei, J., Zheng, B., He, L., Jin, F., 2013. Arithmetic and experimental study of fix-length compensation based on conical bottom shape of electrode in micro-edm. *ASME 2013 International Mechanical Engineering Congress and Exposition*, California.

Pei, J., Zhou, Z., Zhang, L., Zhuang, X., Wu, S., Zhu, Y., Qian, J., 2016. Research on the equivalent plane machining with fix-length compensation method in micro-EDM. *Procedia CIRP* 42, 644–649.

Pham, D.T., Ivanov, A., Bigot, S., Popov, K., Dimov, S., 2007. An investigation of tube and rod electrode wear in micro EDM drilling. *Int. J. Adv. Manuf. Technol.* 33, 103–109.

Puthumana, G., Bissacco, G., Hansen, H.N., 2017. Modeling of the effect of tool wear per discharge estimation error on the depth of machined cavities in micro-EDM milling. *Int. J. Adv. Manuf. Technol.* 92, 3253–3264. doi: 10.1007/s00170-017-0371-5.

Shukla, V., Akhtar, S.N., Subbu, S.K., Ramkumar, J., 2013. fabrication of complex micro channels by micro electric discharge milling. *ASME 2013 International Mechanical Engineering Congress and Exposition*, California, V02BT02A072.

Suzuki, K., Iwai, M., Sharma, A., Sano, S., Uematsu, T., 2006. Low-wear diamond electrode for micro-EDM of die-steel. *Int. J. Manuf. Technol. Manag.* 9, 94–108.

Tong, H., Li, Y., Zhang, L., 2016. On-machine process of rough-and-finishing servo scanning EDM for 3D micro cavities. *Int. J. Adv. Manuf. Technol.* 82, 1007–1015.

Tong, H., Zhang, L., Li, Y., 2014. Algorithms and machining experiments to reduce depth errors in servo scanning 3D micro EDM. *Precis. Eng.* 38, 538–547.

Venugopal, T.R., Muralidhara, R., Rathnamala, R., 2014. Development of micro-EDM incorporating in-situ measurement system. *Procedia Mater. Sci.* 5, 1897–1905.

Wang, J., Qian, J., Ferraris, E., Reynaerts, D., 2017. In-situ process monitoring and adaptive control for precision micro-EDM cavity milling. *Precis. Eng.* 47, 261–275.

Wang, Z., Dong, Y., 2007. Micro-EDM milling of micro compressor prototype. *2007 First International Conference on Integration and Commercialization of Micro and Nanosystems*, Sanya, China. American Society of Mechanical Engineers, 1243–1247.

Yan, M.T., Huang, K.Y., Lo, C.Y., 2009. A study on electrode wear sensing and compensation in micro-EDM using machine vision system. *Int. J. Adv. Manuf. Technol.* 42, 1065–1073.

Yan, M.T., Lin, S.S., 2011. Process planning and electrode wear compensation for 3D micro-EDM. *Int. J. Adv. Manuf. Technol.* 53, 209–219. doi: 10.1007/s00170-010-2827-8.

Yeo, S.H., Kurnia, W., Tan, P.C., Mushan, M., 2007. Development of in situ monitoring and control of micro-EDM process. *Proceedings of the 35th International MATADOR Conference*, Taipei, Taiwan. Springer, 81–84.

Yi, S.M., Park, M.S., Lee, Y.S., Chu, C.N., 2008. Fabrication of a stainless steel shadow mask using batch mode micro-EDM. *Microsyst. Technol.* 14, 411–417.

Yu, H.L., Luan, J.J., Li, J.Z., Zhang, Y.S., Yu, Z.Y., Guo, D.M., 2010. A new electrode wear compensation method for improving performance in 3D micro EDM milling. *J. Micromech. Microeng.* 20, 055011.

Yu, Z.Y., Masuzawa, T., Fujino, M., 1998a. 3D micro-EDM with simple shape electrode. *Int. J. Electr. Mach* 3, 7–12.

Yu, Z.Y., Masuzawa, T., Fujino, M., 1998b. Micro-EDM for three-dimensional cavities - development of uniform wear method. *CIRP Ann.* 47, 169–172.

Yuangang, W., Fuling, Z., Jin, W., 2009. Wear-resist electrodes for micro-EDM. *Chin. J. Aeronaut.* 22, 339–342.

Zhang, L., Du, J., Zhuang, X., Wang, Z., Pei, J., 2015. Geometric prediction of conic tool in micro-EDM milling with fix-length compensation using simulation. *Int. J. Mach. Tools Manuf.* 89, 86–94.

Zhang, L., Jia, Z., Liu, W., Wei, L., 2012. A study of electrode compensation model improvement in micro-electrical discharge machining milling based on large monolayer thickness. *Proc. Inst. Mech. Eng. Part B J. Eng. Manuf.* 226, 789–802.

Index

Printed in the United States
by Baker & Taylor Publisher Services